I0479975

Cosmic Odyssey: A Journey Through the Wonders
of Astronomy

While every precaution has been taken in the preparation of this book, the publisher assumes no responsibility for errors or omissions, or for damages resulting from the use of the information contained herein. COSMIC ODYSSEY: A JOURNEY THROUGH THE WONDERS OF ASTRONOMY

First edition. March 11, 2023. Copyright © 2023 Kenneth Caraballo.

ISBN: 9798388437907

Written by Kenneth Caraballo.

Book Description: Cosmic Odyssey is a fascinating journey through the wonders of astronomy, from the earliest observations of the stars to the latest discoveries about the universe. This book is designed for readers who want to explore the cosmos and learn about the science behind the celestial objects that populate our sky.

The book covers a wide range of topics, including the history of astronomy, the birth and death of stars, the mysteries of black holes and dark matter, the search for extraterrestrial life, and much more.

Starting with the ancient Greeks and their observations of the stars, Cosmic Odyssey takes readers on a journey through the history of astronomy, including the groundbreaking work of Galileo, Kepler, and Newton. The book then dives into the modern era, exploring the latest discoveries made possible by space telescopes and robotic missions.

Throughout the book, readers will learn about the different types of stars, the life cycle of a star, the structure of our galaxy, and the big questions that astronomers are still trying to answer. Cosmic Odyssey also includes fascinating information about the search for extraterrestrial life, the possibility of colonizing other planets, and the future of space exploration.

Written in a clear and engaging style, Cosmic Odyssey is a perfect introduction to astronomy for anyone who is curious about the universe. With its stunning images, fascinating stories, and informative graphics, this book is sure to inspire readers to look up at the night sky with a new sense of wonder and awe.

Chapter 1: Introduction to Astronomy - A Brief History of Astronomy and its Importance in Our World Today

Astronomy is the study of the universe beyond Earth, including the stars, planets, galaxies, and everything in between. It is one of the oldest sciences, with a rich history that stretches back thousands of years.

In ancient times, astronomy played a vital role in human society. Many early civilizations, such as the Babylonians, Egyptians, and Greeks, used the stars and planets to mark the passage of time, to navigate, and to predict events such as eclipses and the changing of the seasons. They also used astronomy to create calendars and to develop religious and mythological beliefs.

One of the most famous astronomical achievements of ancient times was the construction of Stonehenge, a prehistoric monument in England. The stones of Stonehenge were carefully arranged to align with the movements of the Sun and the Moon, making it a kind of astronomical observatory.

In the Middle Ages, astronomy became more closely tied to the field of astrology, which claimed that the positions and movements of the stars and planets could influence human affairs. This led to the development of complex systems for mapping the positions of the stars and planets, as well as the creation of intricate horoscopes.

During the Renaissance, astronomers began to make more precise measurements of the positions of the stars and planets. One of the most famous astronomers of this time was Galileo Galilei, who made many important discoveries using the newly invented telescope. He observed the phases of Venus, the moons of Jupiter, and the rough surface of the Moon, among other things. Galileo's discoveries helped to establish the heliocentric model of the solar system, which placed the Sun at the center instead of the Earth.

The 17th and 18th centuries saw further advances in astronomy. Isaac Newton developed his laws of motion and gravity, which explained the orbits of the planets and the motions of the stars. Astronomers also began to use more powerful telescopes to observe the heavens in greater detail.

In the 20th century, astronomy took a giant leap forward with the development of new technologies such as radio telescopes and space telescopes. These allowed astronomers to study the universe in new ways and to make many groundbreaking discoveries. For example, the discovery of cosmic microwave background radiation provided evidence for the Big Bang theory of the origin of the universe.

Today, astronomy continues to be an important field of study. It has applications in fields as diverse as navigation, timekeeping, communication, and satellite technology. Astronomy also has a profound impact on our understanding of our place in the universe and our origins as a species. It raises fundamental questions about the nature of space and time, the origin and fate of the universe, and the possibility of life on other planets.

In conclusion, astronomy has a long and fascinating history that has played a significant role in human culture and society. Today, it remains a vital field of study that continues to yield new discoveries and insights into the workings of the universe. By studying the stars and planets, we can deepen our understanding of our place in the cosmos and the mysteries that lie beyond.

Chapter 2: The Night Sky - An Overview of the Stars and Constellations Visible in the Night Sky

The night sky is a source of wonder and inspiration for people all over the world. With the naked eye, we can see thousands of stars, as well as planets, galaxies, and other celestial objects. In this chapter, we will explore the stars and constellations visible in the night sky and learn how to identify them.

Stars are massive balls of hot gas that generate energy through nuclear fusion in their cores. They are classified based on their color and temperature, which range from blue-hot to red-cool. The brightest stars in the night sky are often referred to as "first magnitude" stars, while the faintest are "sixth magnitude" stars.

One of the easiest ways to navigate the night sky is by using the constellations. A constellation is a group of stars that appear to form a pattern in the sky. There are 88 official constellations, each with its own unique history and mythology.

Some of the most recognizable constellations in the northern hemisphere include:

- **Ursa Major (the Great Bear):** This constellation is also known as the Big Dipper, which is formed by its seven brightest stars. Ursa Major is circumpolar, which means it never sets below the horizon and is visible year-round in the northern hemisphere.

- **Orion (the Hunter):** This constellation is easily recognizable by its three bright stars in a row that form Orion's Belt. It is also home to the famous Orion Nebula, a cloud of gas and dust where new stars are born.
- **Cassiopeia (the Queen):** This constellation is shaped like a "W" or "M" and is named after the mythical queen of Ethiopia. It is also circumpolar and is visible year-round in the northern hemisphere.
- **Leo (the Lion):** This constellation is named after the lion from Greek mythology and is home to the bright star Regulus.

In the southern hemisphere, some of the most recognizable constellations include:

- **Crux (the Southern Cross):** This constellation is a symbol of the southern hemisphere and is featured on the flags of several countries. It is composed of four bright stars that form a cross.
- **Scorpius (the Scorpion):** This constellation is shaped like a scorpion and is home to the bright star Antares.
- **Centaurus (the Centaur):** This constellation is named after the half-human, half-horse creatures from Greek mythology. It is home to the closest star system to our solar system, Alpha Centauri.

In addition to the constellations, there are also many individual stars that are visible in the night sky. Some of the brightest and most famous stars include Sirius (the brightest star in the sky), Vega (one of the brightest stars in the northern hemisphere), and Betelgeuse (a red supergiant star that is nearing the end of its life).

In conclusion, the night sky is a vast and beautiful expanse that is accessible to anyone with a clear view of the heavens. By learning to identify the stars and constellations, we can deepen our appreciation of the natural world and our place in the universe. Whether you are an amateur astronomer or simply enjoy gazing at the stars, the night sky offers endless possibilities for exploration and discovery.

Chapter 3: Observing the Sky - Tips for Observing the Sky with Your Eyes, Binoculars, and Telescopes

Observing the sky can be a rewarding and exciting experience for anyone interested in astronomy. Whether you're looking at the stars with your naked eye, using binoculars, or a telescope, there are many things to see and explore. In this chapter, we'll provide tips and guidance on how to observe the sky using different tools.

Observing with Your Eyes

Observing the sky with your naked eye is the simplest and most accessible way to explore the night sky. Here are some tips for making the most of your viewing experience:

1. Find a dark location: The darker the location, the better the viewing experience. Avoid light pollution by finding a spot away from cities and towns.
2. Get comfortable: Bring a comfortable chair or blanket to sit on and dress appropriately for the weather.
3. Allow your eyes to adjust: It takes about 20 minutes for your eyes to fully adjust to the darkness, so be patient.
4. Look up: Scan the sky for stars and constellations, and use a star chart or app to help identify what you're seeing.
5. Take breaks: Give your eyes a break by looking away from the sky every once in a while.

Observing with Binoculars

Binoculars can be a great tool for observing the sky. They provide a wider field of view than telescopes, making it easier to locate objects. Here are some tips for observing the sky with binoculars:

1. **Use a tripod:** To avoid shaky images, use a tripod to steady your binoculars.
2. **Choose the right magnification:** Binoculars come in a range of magnifications, but for observing the sky, a magnification of 7x to 10x is recommended.
3. **Focus properly:** Adjust the focus until the image is sharp and clear.
4. **Look for star clusters and nebulae:** Binoculars are great for observing star clusters and nebulae, which can appear as faint smudges of light.

Observing with a Telescope

Telescopes can provide a closer and more detailed look at the sky, but they can also be more complex to use than binoculars. Here are some tips for observing the sky with a telescope:

1. **Choose the right telescope:** Telescopes come in a range of sizes and types, including refracting telescopes and reflecting telescopes. Choose the one that best fits your needs and experience level.

2. **<u>Allow time for set up</u>:** Setting up a telescope can take time, so be patient and follow the instructions carefully.
3. **<u>Use the right eyepiece</u>:** The right eyepiece can make all the difference in the quality of your viewing experience.
4. **<u>Start with easy targets</u>:** Planets like Jupiter and Saturn, as well as the moon, are good places to start when observing with a telescope.
5. **<u>Practice makes perfect</u>:** Like any skill, observing the sky with a telescope takes practice. Don't be discouraged if you don't get it right the first time.

In conclusion, observing the sky can be a fun and educational experience for anyone interested in astronomy. Whether you're using your naked eye, binoculars, or a telescope, there are many objects to see and explore. By following these tips, you can make the most of your viewing experience and deepen your appreciation of the wonders of the night sky.

Chapter 4: Light and Telescopes - Understanding How Telescopes Work and the Different Types Available

Telescopes are essential tools for observing the night sky, but how do they work? In this chapter, we'll explore the science behind telescopes, including how they collect and focus light. We'll also take a look at the different types of telescopes available and their advantages and disadvantages.

The Nature of Light

Light is a form of electromagnetic radiation that travels through space as waves. These waves have a range of wavelengths, and the human eye can only detect a small portion of them, known as the visible spectrum. Telescopes allow us to collect and study light from beyond what the human eye can see, including infrared and ultraviolet light.

Telescopes and Collecting Light

Telescopes collect light and bring it to a focus point, where it can be observed or recorded. The amount of light a telescope can collect is determined by its aperture, which is the diameter of its main mirror or lens. The larger the aperture, the more light the telescope can gather, allowing for clearer and more detailed observations.

Telescope Types

There are three main types of telescopes: refracting, reflecting, and catadioptric. Each has its own advantages and disadvantages.

Refracting Telescopes

Refracting telescopes use lenses to collect and focus light. They were the first type of telescope invented, and are often used for terrestrial viewing as well as astronomy. Refracting telescopes have a simple design, making them easy to use and maintain. However, they suffer from chromatic aberration, which is caused by different wavelengths of light bending at different angles as they pass through the lens, leading to color fringing around the edges of objects.

Reflecting Telescopes

Reflecting telescopes use mirrors to collect and focus light. They were first invented in the 17th century by Sir Isaac Newton, and are now the most commonly used type of telescope in astronomy. Reflecting telescopes do not suffer from chromatic aberration, and can be made with larger apertures than refracting telescopes. However, they require regular maintenance and collimation to ensure that the mirrors are properly aligned.

Catadioptric Telescopes

Catadioptric telescopes use a combination of lenses and mirrors to collect and focus light. They were first invented in the 20th century, and are a hybrid of refracting and reflecting telescopes.

Catadioptric telescopes have a compact design and can provide high-quality images. However, they are more complex than other types of telescopes, making them more difficult to use and maintain.

Conclusion

Telescopes have revolutionized our understanding of the universe, allowing us to study objects beyond what the human eye can see. By understanding how telescopes work and the different types available, we can choose the best instrument for our observing needs. Whether we're observing with a refracting, reflecting, or catadioptric telescope, we can be sure that we are using a tool that has changed our understanding of the cosmos.

Chapter 5: The Solar System - An Overview of the Planets, Moons, Asteroids, and Comets in Our Solar System

The solar system is home to a vast array of celestial objects, including planets, moons, asteroids, and comets. In this chapter, we'll explore each of these objects in detail, as well as their characteristics, formation, and significance.

The Planets

The solar system has eight planets: Mercury, Venus, Earth, Mars, Jupiter, Saturn, Uranus, and Neptune. These planets range in size, composition, and atmosphere. The four inner planets, also known as the terrestrial planets, are small, rocky, and have thin atmospheres. The four outer planets, also known as the gas giants, are much larger and have thick atmospheres made mostly of hydrogen and helium.

Each planet has its own unique features, such as Mars' rust-colored surface and Valles Marineris, the largest canyon in the solar system. Jupiter, the largest planet, has a giant red spot, a storm larger than the size of the Earth, while Saturn has beautiful rings made of ice and dust.

The Moons

Many of the planets in our solar system have moons, which are natural satellites that orbit the planet. The Earth has one moon, while Jupiter has over 80 known moons. These moons come in a variety of sizes, shapes, and compositions. Some are rocky and barren, while others are covered in ice or have active volcanoes.

Some of the most interesting moons in our solar system include Io, a volcanic moon of Jupiter, and Enceladus, a moon of Saturn with an underground ocean that could potentially harbor life.

Asteroids and Comets

Asteroids and comets are small, rocky or icy objects that orbit the Sun. Asteroids are typically found in the asteroid belt, between Mars and Jupiter, while comets are found in the outer reaches of the solar system, beyond the orbit of Neptune.

Asteroids range in size from tiny pebbles to large rocks, and some are large enough to be classified as dwarf planets. Comets are made up of ice, dust, and rock, and often have long tails that are visible from Earth as they approach the Sun.

Significance of the Solar System

The study of the solar system is significant for several reasons. By studying the formation and evolution of the planets, we can learn about the history of the solar system and the conditions that allowed life to arise on Earth. The study of asteroids and comets can also help us understand the risks of impact events, which could potentially cause catastrophic damage to Earth.

Conclusion

The solar system is a fascinating and complex system of planets, moons, asteroids, and comets. Each object has its own unique characteristics and significance, providing us with valuable insights into the formation and evolution of the universe. By studying the solar system, we can deepen our understanding of our place in the universe and the potential for life beyond Earth.

Chapter 6: The Sun - The Life Cycle of Our Nearest Star and Its Importance to Life on Earth

The Sun is the closest star to Earth and is essential to life on our planet. In this chapter, we'll explore the life cycle of the Sun, its structure, and its significance to our planet.

Structure of the Sun

The Sun is a massive ball of hot plasma, which is a state of matter made up of charged particles. It is composed mostly of hydrogen and helium, with trace amounts of other elements. At the center of the Sun is the core, where nuclear fusion reactions occur, producing immense amounts of energy.

Surrounding the core is the radiative zone, where energy is transported outward by radiation. Beyond the radiative zone is the convective zone, where energy is transported outward by convection. The outermost layer of the Sun is the atmosphere, which includes the photosphere, the chromosphere, and the corona.

Life Cycle of the Sun

The Sun has a life cycle that spans billions of years. It began as a giant cloud of gas and dust, called a nebula, which collapsed under its own gravity to form the protostar that eventually became the Sun.

The Sun is currently in the main sequence stage of its life cycle, where it is fusing hydrogen into helium in its core. This process releases a tremendous amount of energy, which radiates outward and provides the heat and light that sustains life on Earth.

Eventually, the Sun will exhaust the hydrogen in its core and begin fusing helium into heavier elements. This will cause the Sun to expand and become a red giant, engulfing the inner planets, including Earth. As the red giant stage ends, the Sun will shed its outer layers, leaving behind a hot, dense core known as a white dwarf.

Importance of the Sun to Life on Earth

The Sun is essential to life on Earth. Its energy provides the heat and light needed for photosynthesis, the process by which plants convert sunlight into energy. This energy is then transferred through the food chain, sustaining life on our planet.

The Sun also drives our planet's weather and climate patterns. Its energy heats the Earth's atmosphere, creating winds and ocean currents that circulate heat around the planet.

However, the Sun's energy can also be harmful. The Sun emits ultraviolet radiation, which can cause skin cancer and damage to the ozone layer. Solar flares and coronal mass ejections can also have a significant impact on our planet's technology and power grids.

Conclusion

The Sun is a vital component of our solar system and essential to life on Earth. Its life cycle and structure provide us with valuable insights into the formation and evolution of our universe. While the Sun's energy is necessary for life, it can also have harmful effects. Understanding the Sun and its impact on our planet is crucial for our continued existence and progress as a species.

Chapter 7: The Moon - The Formation and Features of Earth's Only Natural Satellite

The Moon is Earth's only natural satellite and has fascinated humans for centuries. In this chapter, we'll explore the formation of the Moon, its features, and its significance to Earth.

Formation of the Moon

The current theory for the Moon's formation is the giant impact hypothesis. This hypothesis suggests that a Mars-sized object collided with Earth about 4.5 billion years ago. The impact caused debris to be ejected into space, which eventually coalesced into the Moon.

The evidence for this theory includes the similarity in chemical composition between the Moon and Earth, as well as the fact that the Moon's orbit is in the same plane as Earth's equator.

Features of the Moon

The Moon's surface is characterized by impact craters, mountains, valleys, and flat plains known as maria. The maria are believed to have been formed by volcanic activity, while the craters and mountains were formed by impacts from meteoroids and asteroids.

The Moon has no atmosphere or magnetic field, which means it is exposed to solar radiation and cosmic rays. The temperature on the Moon varies greatly, from over 100 degrees Celsius during the day to minus 170 degrees Celsius at night.

The Moon's surface is also covered in a fine layer of dust known as regolith, which was formed by the constant bombardment of meteoroids and asteroids. The regolith has been extensively studied by astronauts and robotic missions, providing valuable insights into the Moon's history and formation.

Significance of the Moon to Earth

The Moon has a significant impact on Earth. Its gravitational pull creates tides in our oceans, which can affect navigation, fishing, and coastal erosion. The Moon also stabilizes Earth's rotation, preventing it from wobbling on its axis.

The Moon has been an object of human fascination for centuries and has inspired countless works of art, literature, and science. It has also been the site of several manned and unmanned missions, including the Apollo missions, which brought humans to its surface for the first time.

Conclusion

The Moon is a fascinating object that has captured the human imagination for centuries. Its formation and features provide valuable insights into the history and formation of our solar system. The Moon's impact on Earth, from its gravitational pull to its cultural significance, is significant and deserves further study and exploration.

Chapter 8: Mercury - A Closer Look at the Closest Planet to the Sun

Mercury is the smallest planet in our solar system and also the closest to the Sun. In this chapter, we will explore the characteristics of Mercury, its composition, and the challenges of studying this extreme planet.

Characteristics of Mercury

Mercury is a rocky, terrestrial planet with a diameter of just over 4,800 kilometers, making it only slightly larger than Earth's Moon. It has a very thin atmosphere, which means that it has no significant weather or seasons.

The temperature on Mercury is extreme, ranging from over 400 degrees Celsius during the day to minus 170 degrees Celsius at night. This is due to its proximity to the Sun and its lack of an atmosphere to regulate the temperature.

Mercury also has a slow rotation, taking almost 59 Earth days to complete one rotation on its axis. However, its orbit around the Sun is much faster, taking only 88 Earth days.

Composition of Mercury

Mercury's composition is similar to that of Earth's Moon, consisting of a large iron core, a thin mantle, and a rocky crust. The core of Mercury is relatively large, making up about 42% of its volume.

The surface of Mercury is covered in craters, similar to the Moon, and has a range of features including valleys, ridges, and cliffs. The planet also has large flat plains, known as "maria," which were likely formed by volcanic activity.

Studying Mercury

Studying Mercury is challenging due to its proximity to the Sun, which means that spacecraft must be designed to withstand extreme heat and radiation. To date, there have been only two missions to Mercury: NASA's Mariner 10 in 1974 and the European Space Agency's BepiColombo mission, which launched in 2018.

Both missions provided valuable insights into Mercury's characteristics and composition. Mariner 10 imaged only about 45% of the planet's surface, while the BepiColombo mission aims to provide a more comprehensive understanding of the planet.

Conclusion

Mercury is an extreme planet with unique characteristics that make it challenging to study. Despite its proximity to the Sun, it has a surprising amount in common with Earth's Moon in terms of its composition and surface features. Further study of this planet will provide valuable insights into the formation and history of our solar system.

Chapter 9: Venus - The Hottest Planet in the Solar System and Its Unique Features

Venus, also known as the Morning Star or the Evening Star, is the second planet from the Sun and the hottest planet in our solar system. In this chapter, we will explore the characteristics of Venus, its composition, and its unique features.

Characteristics of Venus

Venus is similar in size and composition to Earth, and is often referred to as Earth's sister planet. However, it has a very different atmosphere and surface characteristics. The surface of Venus is shrouded in thick clouds that reflect sunlight, giving the planet a bright, white appearance. The atmosphere is mostly composed of carbon dioxide, with small amounts of nitrogen and other gasses.

The temperature on the surface of Venus is extremely hot, with an average temperature of about 462 degrees Celsius, making it the hottest planet in the solar system. The high temperature is due to the greenhouse effect caused by the thick atmosphere, which traps heat and prevents it from escaping into space.

Composition of Venus

Venus has a rocky, terrestrial composition similar to Earth, with a large iron core, a mantle, and a rocky crust. The atmosphere of

Venus is very dense and has a pressure that is over 90 times greater than that of Earth's atmosphere at sea level.

Unique Features of Venus

One of the most unique features of Venus is its retrograde rotation, meaning it rotates on its axis in the opposite direction to the other planets in the solar system. The exact cause of this unusual rotation is still not well understood.

Venus also has no moons or rings and its magnetic field is much weaker than Earth's. The planet is covered in volcanic features, including large shield volcanoes and vast lava plains.

Studying Venus

Studying Venus is challenging due to its thick atmosphere, which makes it difficult to observe the planet's surface. However, several spacecraft have been sent to study Venus, including NASA's Magellan spacecraft and the European Space Agency's Venus Express mission.

The Magellan spacecraft used radar to map the surface of Venus, revealing detailed images of its volcanoes, mountains, and valleys. The Venus Express mission focused on studying the planet's atmosphere, revealing new insights into the composition and dynamics of its thick, carbon dioxide-rich atmosphere.

Conclusion

Venus is a fascinating planet with unique characteristics and a challenging environment for study. Its thick atmosphere and high temperature make it one of the most extreme planets in the solar system. Further study of this planet will provide valuable insights into the formation and evolution of terrestrial planets and the conditions necessary for life to exist in the universe.

Chapter 10: Mars - The Red Planet, Its History, and the Ongoing Exploration of Its Surface

Mars, also known as the Red Planet, has captivated humans for centuries with its distinctive color and proximity to Earth. In recent years, a number of missions have been launched to study Mars, uncovering new insights about the planet's geology, climate, and potential for life. In this chapter, we will explore the history of Mars, its characteristics, and the ongoing exploration of its surface.

History of Mars

Mars is the fourth planet from the Sun and the second smallest planet in our solar system. It is named after the Roman god of war due to its red appearance, caused by iron oxide (rust) on its surface. Mars has been known since ancient times, and has been studied by astronomers for centuries. In the late 19th and early 20th centuries, astronomers began to make observations of the planet's surface, revealing craters, mountains, and valleys.

Characteristics of Mars

Mars has a thin atmosphere, with mostly carbon dioxide and small amounts of nitrogen and argon. The planet's surface is rocky and barren, with a variety of features such as canyons, craters, and mountains. Mars has two small moons, Phobos and Deimos, which are thought to be captured asteroids.

One of the most intriguing characteristics of Mars is its potential for harboring life. While there is no evidence of current life on the planet, scientists have found evidence of water on Mars, suggesting that the planet may have had conditions suitable for life in the past.

Exploration of Mars

The exploration of Mars began in the 1960s with flyby missions by the United States and the Soviet Union. Since then, a number of missions have been launched to study the planet in more detail. These missions include orbiters, landers, and rovers, which have provided valuable information about Mars' geology, atmosphere, and potential for life.

One of the most successful Mars missions was NASA's Mars Exploration Rover mission, which landed two rovers, Spirit and Opportunity, on the planet's surface in 2004. The rovers explored the Martian surface for several years, studying rocks and soil and providing valuable information about the planet's geology and history.

In recent years, several new missions have been launched to study Mars. These include NASA's Mars Science Laboratory mission, which landed the Curiosity rover on Mars in 2012, and the Mars 2020 mission, which landed the Perseverance rover on the planet's surface in February 2021. These missions are focused on studying the potential for life on Mars, with a particular focus on searching for signs of ancient microbial life.

Conclusion

Mars is a fascinating planet with a rich history and a promising future for exploration. The ongoing missions to study the planet are providing valuable insights into its geology, climate, and potential for life. As we continue to explore Mars, we may uncover new clues about the formation and evolution of our solar system, and the conditions necessary for life to exist beyond Earth.

Chapter 11: Jupiter - The Largest Planet in the Solar System and Its Fascinating Moons

Jupiter is the fifth planet from the Sun and the largest planet in the solar system, with a diameter of over 86,000 miles. It is a gas giant, composed mostly of hydrogen and helium, and is known for its distinctive stripes and Great Red Spot, a massive storm that has been raging for over 350 years. In this chapter, we will explore the characteristics of Jupiter, its moons, and the ongoing missions to study this fascinating planet.

Characteristics of Jupiter

Jupiter is a gas giant, meaning that it does not have a solid surface. Instead, its atmosphere gradually gets denser and eventually becomes a liquid. The planet's atmosphere is composed mostly of hydrogen and helium, with small amounts of other elements such as methane, ammonia, and water vapor.

One of the most striking features of Jupiter is its colorful bands and zones, created by strong winds that blow in opposite directions. The planet's Great Red Spot, a giant storm that is larger than the size of Earth, is also a famous feature of Jupiter's atmosphere. The storm has been raging for over 350 years and is thought to be a high-pressure system that creates a vortex.

Jupiter's Moons

Jupiter has 79 known moons, the largest of which are the four Galilean moons: Io, Europa, Ganymede, and Callisto. These moons were discovered by Galileo Galilei in 1610 and are named after him. Each of the Galilean moons is unique, with its own distinct characteristics and features.

Io is the most volcanically active object in the solar system, with over 400 active volcanoes. Its surface is covered in sulfur and sulfur dioxide, giving it a yellow and orange color.

Europa is one of the most intriguing moons in the solar system, as it is believed to have a subsurface ocean of liquid water. This has led to speculation that Europa may harbor life.

Ganymede is the largest moon in the solar system and is the only moon known to have its own magnetic field. Its surface is covered in craters, as well as a network of valleys and ridges.

Callisto is the third-largest moon in the solar system and is the most heavily cratered. It is also the furthest from Jupiter of the Galilean moons.

Exploration of Jupiter

The exploration of Jupiter began in the 1970s with the Pioneer and Voyager missions, which flew by the planet and provided the first close-up images of its atmosphere and moons. In 1995, the Galileo spacecraft was launched to study Jupiter in more detail. It orbited the planet for over seven years and provided detailed images of its atmosphere and moons.

In 2016, NASA's Juno spacecraft arrived at Jupiter to study its atmosphere, magnetic field, and interior structure. Juno has provided new insights into Jupiter's deep atmosphere, revealing that its belts and zones extend far deeper than previously thought. The spacecraft has also provided detailed images of Jupiter's polar regions, which were previously unseen.

Conclusion

Jupiter is a fascinating planet with a complex atmosphere and a diverse system of moons. The ongoing missions to study Jupiter are providing valuable insights into the planet's composition, history, and potential for life. As we continue to explore Jupiter and its moons, we may uncover new clues about the formation and evolution of our solar system, and the conditions necessary for life to exist beyond Earth.

Chapter 12: Saturn - The Ringed Planet and Its Complex System of Rings and Moons

Saturn is the sixth planet from the Sun and the second-largest planet in the solar system, after Jupiter. It is a gas giant, similar to Jupiter, but is best known for its beautiful and unique system of rings. In this chapter, we will explore the characteristics of Saturn, its rings, and its fascinating system of moons.

Characteristics of Saturn

Saturn is a gas giant, composed mostly of hydrogen and helium, with small amounts of other elements such as methane and ammonia. Its atmosphere is similar to that of Jupiter, with colorful bands and zones created by strong winds.

Saturn has a prominent feature that sets it apart from all other planets in the solar system: its complex system of rings. These rings are made up of millions of individual particles of ice and rock, ranging in size from tiny grains to larger boulders. The rings are divided into several distinct regions, each with its own unique characteristics.

Saturn's Moons

Saturn has at least 82 known moons, the largest of which are Titan, Rhea, Enceladus, Dione, and Iapetus. Titan is the largest moon of Saturn and the second-largest moon in the solar system, after Jupiter's Ganymede. It is the only moon in the solar system with a thick atmosphere, composed mostly of nitrogen, and also has lakes and seas of liquid methane and ethane on its surface.

Enceladus is another fascinating moon of Saturn, as it is believed to have a subsurface ocean of liquid water. It is also known for its geysers, which shoot plumes of water vapor and ice particles into space.

Rhea, Dione, and Iapetus are also important moons of Saturn, with their own unique features and characteristics. Rhea has a heavily cratered surface, while Dione has a complex system of ridges and valleys. Iapetus is known for its unusual two-tone coloration, with one hemisphere being much darker than the other.

Saturn's Rings

Saturn's rings are made up of millions of individual particles, ranging in size from tiny grains to larger boulders. The rings are divided into several distinct regions, each with its own unique characteristics. The outermost region of the rings, known as the E ring, is the largest and most diffuse, and is believed to be created by material ejected from Enceladus's geysers.

The rings are believed to be relatively young, possibly only a few hundred million years old, and are constantly changing. The particles in the rings collide with each other, creating new fragments and changing the overall shape of the rings.

Exploration of Saturn

The exploration of Saturn began in the 1970s with the Pioneer and Voyager missions, which provided the first close-up images of the planet and its rings. In 2004, the Cassini spacecraft was launched to study Saturn in more detail. Cassini orbited the planet for over 13 years and provided detailed images of its atmosphere, rings, and moons.

Cassini also discovered new moons of Saturn and provided detailed images of the planet's largest moon, Titan. In 2017, the spacecraft intentionally crashed into Saturn's atmosphere, bringing an end to its mission.

Conclusion

Saturn is a fascinating planet with a unique system of rings and a diverse system of moons. The ongoing missions to study Saturn are providing valuable insights into the planet's composition, history, and potential for life. As we continue to explore Saturn and its moons, we may uncover new clues about the formation and evolution of our solar system, and the conditions necessary for life to exist beyond Earth.

Chapter 13: Uranus - A closer look at the ice giant planet and its unique tilt.

Uranus is the seventh planet from the Sun and the third-largest planet in our solar system. It is classified as an ice giant planet because it is primarily composed of various ices, such as water, methane, and ammonia. It is also unique among the planets in our solar system due to its extreme tilt.

Discovery and Exploration:

Uranus was discovered in 1781 by Sir William Herschel, a British astronomer. Herschel initially believed he had found a comet, but further observations revealed that it was actually a planet. Herschel named the planet after the Greek god of the sky, Uranus.

The first spacecraft to visit Uranus was Voyager 2 in 1986. Voyager 2 provided the first detailed images of the planet and its moons. It also revealed Uranus's unique magnetic field and the extreme tilt of its axis.

Physical Characteristics:

Uranus has a diameter of approximately 51,118 km and a mass roughly 14.5 times that of Earth. It is the coldest planet in our solar system, with an average temperature of -195 degrees Celsius.

Uranus is primarily composed of hydrogen and helium, like Jupiter and Saturn. However, it also contains a high proportion of ices, such as water, methane, and ammonia. These ices give Uranus its blue-green color.

One of the most unusual features of Uranus is its extreme tilt. The planet's axis of rotation is tilted at an angle of 98 degrees, which means that its poles are almost in the plane of its orbit. This tilt is thought to be the result of a collision with a massive object early in Uranus's history.

Moons:

Uranus has 27 known moons, all of which are named after characters from the works of William Shakespeare and Alexander Pope. The five largest moons are Miranda, Ariel, Umbriel, Titania, and Oberon.

Miranda is the smallest of the five largest moons and has one of the most varied and interesting landscapes of any moon in our solar system. Ariel is the brightest and most reflective of Uranus's moons. Umbriel is the darkest of the five largest moons and has a heavily cratered surface. Titania and Oberon are both heavily cratered, but also have many large canyons and fault systems.

Conclusion:

Uranus is a fascinating planet with many unique features, including its extreme tilt and composition. Although it has only been visited by one spacecraft to date, future missions may reveal more about this distant ice giant planet and its system of moons.

Chapter 14: Neptune - The farthest planet from the Sun and its mysterious Great Dark Spot.

Neptune is the eighth and farthest planet from the Sun in our solar system. It is classified as an ice giant planet, like Uranus, and was discovered in 1846 by the French astronomer Urbain Le Verrier. Neptune is the fourth-largest planet in our solar system, and it is named after the Roman god of the sea.

Physical Characteristics:

Neptune has a diameter of approximately 49,244 km and a mass about 17 times that of Earth. It is similar in composition to Uranus, being made primarily of hydrogen and helium with a significant amount of ices, including water, ammonia, and methane. These ices give Neptune its blue color.

The average temperature on Neptune is approximately -214 degrees Celsius, making it one of the coldest planets in our solar system. Neptune's atmosphere is primarily composed of hydrogen and helium, with traces of methane, ethane, and other hydrocarbons. The methane in Neptune's atmosphere gives it its characteristic blue color, as it absorbs red light and reflects blue light.

Moons:

Neptune has 14 known moons, the largest of which is Triton. Triton is the only large moon in the solar system that orbits its planet in a retrograde direction, meaning it orbits in the opposite direction to Neptune's rotation. Triton is also unique in that it has a surface temperature of -235 degrees Celsius, making it one of the coldest objects in the solar system.

Great Dark Spot:

Neptune's most notable feature is its Great Dark Spot, which is a massive storm system in its atmosphere. The Great Dark Spot was first observed by the Voyager 2 spacecraft in 1989 and is similar in size to Jupiter's Great Red Spot. The storm system is located in Neptune's southern hemisphere and is thought to be a high-pressure system.

However, observations made by the Hubble Space Telescope in the late 1990s showed that the Great Dark Spot had disappeared, suggesting that Neptune's weather patterns are highly variable and unpredictable. It is thought that the storm system may have dissipated due to changes in Neptune's atmosphere or because it moved to another location on the planet.

Conclusion:

Neptune is a fascinating planet, and despite its distance from the Sun, it has many unique features and mysteries waiting to be uncovered. Its Great Dark Spot is a notable feature and a reminder of the dynamic and ever-changing nature of our solar system. With future missions and advances in technology, we can expect to learn even more about this distant ice giant planet and its system of moons.

Chapter 15: Dwarf Planets and Asteroids - A tour of the smaller bodies in our solar system.

In addition to the eight planets in our solar system, there are many smaller bodies, including dwarf planets and asteroids. These objects provide valuable information about the history and formation of our solar system.

Dwarf Planets:

Dwarf planets are celestial bodies that orbit the Sun but are not large enough to be considered full-fledged planets. In our solar system, there are five recognized dwarf planets: Ceres, Pluto, Haumea, Makemake, and Eris.

Ceres is the smallest dwarf planet and is located in the asteroid belt between Mars and Jupiter. It is the largest object in the asteroid belt and was the first dwarf planet to be discovered. Ceres is thought to be composed of rock and ice, and recent observations by NASA's Dawn spacecraft suggest that it may have a subsurface ocean of liquid water.

Pluto was once considered the ninth planet in our solar system, but in 2006, it was reclassified as a dwarf planet. It is located in the Kuiper Belt, a region of the solar system beyond Neptune, and is smaller than Earth's Moon. Pluto has a highly elliptical orbit and is thought to be composed of rock and ice.

Haumea, Makemake, and Eris are all located in the Kuiper Belt and are similar in size to Pluto. Haumea is notable for its elongated shape, which is thought to be the result of a collision with another object. Makemake is known for its reddish-brown color, and Eris is the largest known dwarf planet and is located in the outer reaches of the solar system.

Asteroids:

Asteroids are small, rocky objects that orbit the Sun, and are remnants from the formation of our solar system. There are millions of asteroids in our solar system, ranging in size from a few meters to several hundred kilometers in diameter.

Most asteroids are located in the asteroid belt between Mars and Jupiter, but some asteroids have orbits that cross the paths of the planets. These asteroids, known as Near-Earth Objects (NEOs), pose a potential hazard to Earth if they were to collide with our planet.

One of the most well-known asteroids is Bennu, which is being studied by NASA's OSIRIS-REx spacecraft. Bennu is thought to be a primitive asteroid, which means it has remained relatively unchanged since the formation of our solar system. The OSIRIS-REx mission aims to collect a sample of Bennu's surface and return it to Earth for study.

Conclusion:

Dwarf planets and asteroids provide valuable information about the history and formation of our solar system. These objects have unique features and characteristics, and studying them can help us better understand the processes that shaped our solar system. With ongoing missions and advances in technology, we can expect to learn even more about these fascinating objects in the future.

Chapter 16: Comets - The Icy Wanderers That Visit Our Solar System from the Depths of Space

Comets are some of the most fascinating objects in our solar system, with their long tails and unpredictable appearances. These icy wanderers come from the Kuiper Belt and Oort Cloud, two regions beyond the orbit of Neptune, and provide scientists with valuable information about the early solar system.

In this chapter, we'll explore what comets are, where they come from, and how they behave. We'll also discuss the impact of comets on our planet and the study of comets by spacecraft.

What Are Comets?

Comets are small, icy bodies that orbit the Sun. They are often referred to as "dirty snowballs" because they are made up of a mixture of ice, dust, and rock. The ice in comets is primarily frozen water, but it can also include frozen gasses like methane, ammonia, and carbon dioxide.

Comets come in two main types: short-period comets and long-period comets. Short-period comets have orbits that take them around the Sun in less than 200 years and are thought to originate from the Kuiper Belt, a region of the solar system beyond the orbit of Neptune. Long-period comets have orbits that take them more than 200 years to orbit the Sun and are believed to come from the Oort Cloud, a region of icy bodies that surrounds the solar system at a distance of up to 100,000 AU.

Comet Anatomy

A comet has two main parts: the nucleus and the coma. The nucleus is the solid, central part of the comet and is made up of rock, dust, and ice. The coma is the fuzzy, cloud-like area that surrounds the nucleus and is made up of gas and dust that have been released from the nucleus as it is heated by the Sun.

As a comet approaches the Sun, the heat causes the ice in the nucleus to sublimate, or turn directly from a solid to a gas, releasing dust and gas into space. This forms the coma and also creates a tail that can extend for millions of miles.

Comet Impact

Comets have had a significant impact on our planet over the course of history. The most famous impact is the one that likely caused the extinction of the dinosaurs 65 million years ago. This impact is thought to have been caused by a comet or asteroid that was approximately 6 miles in diameter.

In addition to causing catastrophic events, comets also bring valuable resources to our planet. The water in comets is believed to be the source of Earth's oceans, and they contain organic molecules that are the building blocks of life.

Spacecraft Exploration

Scientists have been studying comets for decades, and several spacecraft have been sent to observe them up close. The first spacecraft to visit a comet was the European Space Agency's Giotto mission, which flew by Halley's Comet in 1986. NASA's Deep Impact mission, which launched in 2005, impacted a comet to study its interior.

The Rosetta mission, launched by the European Space Agency in 2004, orbited and landed on a comet named 67P/Churyumov-Gerasimenko in 2014. The mission provided scientists with unprecedented data on the composition of comets and helped shed light on their origins.

Conclusion

Comets are fascinating objects that provide insight into the early solar system and the potential for life in our universe. Their unpredictable appearances and long tails make them a spectacle to observe, and their impact on our planet cannot be ignored. As we continue to study comets, we will undoubtedly uncover more about the history and future of our solar system.

Chapter 17: Meteoroids, Meteors, and Meteorites - Understanding the differences between these space rocks.

The universe is full of space rocks, ranging from tiny grains of dust to massive asteroids and comets. As these objects travel through space, they can encounter Earth's atmosphere and create dazzling displays of light and sound known as meteors or shooting stars. In some cases, these space rocks can even survive their fiery descent and reach the ground as meteorites.

In this chapter, we will explore the different types of space rocks and the phenomena they create.

Meteoroids

Meteoroids are small rocks or particles that exist in space. They are the source of most meteors that we see streaking across the night sky. These objects can range in size from a grain of sand to several meters in diameter. Most meteoroids originate from comets or asteroids, which release dust and debris as they orbit the Sun.

Meteors

When a meteoroid enters Earth's atmosphere, it creates a streak of light in the sky known as a meteor. Meteors are sometimes called shooting stars, but they are not stars at all. They are actually the result of the friction between the meteoroid and the air molecules in Earth's atmosphere. This friction causes the meteoroid to heat up and vaporize, creating a trail of glowing gas that we see as a streak of light.

Meteor Showers

Meteor showers occur when Earth passes through a trail of debris left behind by a comet or asteroid. When the Earth encounters this debris, the meteoroids enter the atmosphere at a high rate, creating many meteors in a short period of time. Meteor showers are named after the constellation from which the meteors appear to originate. The Perseid meteor shower, for example, appears to originate from the constellation Perseus.

Fireballs

A fireball is a particularly bright meteor that is visible even in daylight. Fireballs occur when a large meteoroid enters Earth's atmosphere. These objects can produce a flash of light that is brighter than the full Moon.

Meteorites

When a meteoroid survives its fiery descent through Earth's atmosphere and reaches the ground, it becomes a meteorite. Meteorites can provide valuable information about the early solar system, as they are believed to be remnants from the formation of the planets. Scientists study meteorites to learn about the chemical composition of the solar system and to understand the processes that formed our planet.

Conclusion

In conclusion, meteoroids, meteors, and meteorites are all different types of space rocks that can provide a wealth of information about the solar system. While meteoroids and meteors create beautiful displays in the night sky, meteorites allow us to study the building blocks of the solar system up close.

Chapter 18: The Kuiper Belt and Oort Cloud - The distant regions beyond Neptune that contain countless icy bodies.

The Kuiper Belt and Oort Cloud are two of the most fascinating and mysterious regions of our solar system. These distant areas beyond Neptune are home to countless icy bodies, including dwarf planets, comets, and other objects. In this chapter, we will explore the Kuiper Belt and Oort Cloud, their origins, and their significance in our understanding of the solar system.

The Kuiper Belt:

The Kuiper Belt is a region beyond Neptune that is home to a large number of small icy bodies, including dwarf planets like Pluto, Haumea, and Makemake. The region is named after Dutch-American astronomer Gerard Kuiper, who first proposed its existence in the 1950s. Kuiper suggested that beyond the orbit of Neptune, there might be a belt of small icy bodies that were left over from the formation of the solar system.

The Kuiper Belt extends from about 30 to 50 astronomical units (AU) from the Sun, which is roughly 4.5 to 7.5 billion kilometers. The region is believed to contain tens of thousands of objects that are larger than 100 kilometers in diameter, and possibly millions of smaller objects.

One of the most famous objects in the Kuiper Belt is Pluto, which was classified as a planet until 2006 when it was reclassified as a dwarf planet. Pluto is only one of many dwarf planets in the Kuiper Belt, and scientists believe that there are likely dozens more waiting to be discovered.

The Kuiper Belt is believed to be the source of short-period comets, which have orbits that take them around the Sun in less than 200 years. These comets are thought to have been ejected from the Kuiper Belt by the gravitational influence of the gas giant planets, such as Jupiter and Saturn.

The Oort Cloud:

The Oort Cloud is a much more distant region of the solar system, believed to be located at a distance of around 50,000 astronomical units (AU) from the Sun. This is roughly one light-year away from the Sun, and it is so far away that it is essentially impossible to observe directly.

The Oort Cloud is believed to be the source of long-period comets, which have orbits that take them around the Sun in more than 200 years. These comets are thought to have been ejected from the Oort Cloud by the gravitational influence of passing stars or other celestial objects.

The Oort Cloud is named after Dutch astronomer Jan Oort, who first proposed its existence in the 1950s. Oort suggested that the Oort Cloud was a vast spherical shell of icy bodies that surrounded the solar system, and that it was the source of comets that visited the inner solar system.

The significance of the Kuiper Belt and Oort Cloud:

The Kuiper Belt and Oort Cloud are important because they provide clues about the formation and evolution of the solar system. Scientists believe that the icy bodies in these regions are remnants from the early solar system, and studying them can tell us about the conditions and processes that existed during the formation of the planets.

In addition, the Kuiper Belt and Oort Cloud are home to some of the most fascinating and mysterious objects in the solar system, including dwarf planets, comets, and other icy bodies. Studying these objects can help us understand the nature of the early solar system and the processes that have shaped it over time.

Conclusion:

The Kuiper Belt and Oort Cloud are fascinating regions of our solar system that contain countless icy bodies. The discovery of these regions has expanded our understanding of the outer solar system and the formation of the planets. With ongoing advancements in technology and space exploration, there is much more to learn about these distant regions and the objects they contain.

As we continue to explore our solar system and beyond, it is important to remember that the discoveries we make can have a profound impact on our understanding of the universe and our place within it. The study of astronomy allows us to appreciate the beauty and complexity of the cosmos and inspires us to continue pushing the boundaries of our knowledge.

In conclusion, the Kuiper Belt and Oort Cloud are just two of the many wonders of our solar system, and their exploration is an important part of our ongoing quest to understand the universe. Through continued observation and research, we can uncover new insights into the formation and evolution of our solar system and the cosmos as a whole.

Chapter 19: Stars - The different types of stars and how they form and evolve

Stars are the most fundamental and abundant objects in the universe. They come in a range of sizes, colors, and temperatures, and understanding their properties and behavior is key to understanding the universe as a whole. In this chapter, we will explore the different types of stars and how they form and evolve over time.

The Birth of Stars

Stars begin their lives as clouds of gas and dust known as nebulae. Gravity causes these clouds to collapse in on themselves, forming a dense core that eventually becomes hot enough to trigger nuclear fusion. This is the process by which hydrogen atoms are fused together to form helium, releasing energy in the process. This energy generates the heat and light that make stars shine.

As the fusion reactions continue, the star enters what is known as the main sequence phase. This is when a star settles into a state of stable fusion, during which it will burn through its fuel reserves over millions or even billions of years.

Types of Stars

There are many different types of stars, and they can be classified according to their temperature, size, and luminosity.

The most common way to classify stars is by using the Hertzsprung-Russell (HR) diagram, which plots a star's temperature against its luminosity.

The HR diagram shows that stars fall into several distinct groups, including main sequence stars, giant stars, and supergiant stars. Main sequence stars are the most common type of star, and they are characterized by their relatively stable fusion reactions. Giant stars are much larger than main sequence stars, and they have a much higher luminosity. Supergiant stars are the largest and most luminous stars of all, and they often end their lives in spectacular supernova explosions.

Stellar Evolution

As a star burns through its fuel reserves, it begins to evolve and change in appearance. The evolution of a star depends on its mass. Low-mass stars like our Sun will eventually exhaust their fuel and become red giants, which are much larger and cooler than their main sequence counterparts. After shedding their outer layers, these stars will eventually become white dwarfs, which are extremely dense objects about the size of Earth.

High-mass stars, on the other hand, will eventually become supernovae. These massive explosions are among the most energetic events in the universe, and they can briefly outshine entire galaxies. After the explosion, the remaining material can form a neutron star or a black hole.

Conclusion

Studying stars is crucial to our understanding of the universe. By examining their properties and behavior, astronomers can learn about the formation and evolution of galaxies, the structure of the universe, and the origins of life itself. The different types of stars, from main sequence stars to red giants and supernovae, provide us with a wealth of information about the processes that shape the cosmos. As we continue to study stars and other celestial objects, we will continue to uncover new insights into the mysteries of the universe.

Chapter 20: The Hertzsprung-Russell Diagram - A tool for understanding the properties of stars.

Stars are the most fundamental objects in astronomy. They are the building blocks of galaxies, the engines of the universe, and the sources of light and energy that make life on Earth possible. In order to understand stars, astronomers have developed many tools and techniques to study them. One of the most important of these is the Hertzsprung-Russell diagram, also known as the HR diagram.

The HR diagram is a plot of the luminosity (brightness) of stars versus their temperature (color). It was first developed independently by Danish astronomer Ejnar Hertzsprung and American astronomer Henry Norris Russell in the early 20th century. The diagram has since become one of the most important tools in astronomy for studying stars and their properties.

The HR diagram is a scatter plot of data points, with each point representing a star. The x-axis of the plot represents the temperature of the star, usually measured in Kelvin. The y-axis represents the luminosity of the star, usually measured in terms of the Sun's luminosity ($L_\odot$). The Sun's luminosity is about 3.8 x 10^26 watts.

When the data points on the HR diagram are plotted, they form a distinct pattern. The majority of stars fall along a diagonal line called the main sequence. This line represents stars that are fusing hydrogen into helium in their cores, which is the process that powers stars like the Sun. Stars on the main sequence range from small, cool red dwarfs to massive, hot blue giants.

Above and below the main sequence are other groups of stars with different properties. These groups include white dwarfs, red giants, and supergiants. Each of these groups represents stars that have evolved beyond the main sequence and have different properties due to changes in their internal structure.

The HR diagram is a powerful tool because it allows astronomers to determine a star's temperature, luminosity, and other properties based on its position on the diagram. By studying the HR diagram, astronomers have been able to learn a great deal about the properties and behavior of stars.

For example, the HR diagram can be used to estimate a star's age. Since a star's luminosity and temperature change over time as it evolves, its position on the HR diagram can give astronomers an idea of how long it has been burning hydrogen in its core. This information can help astronomers understand how stars form and evolve over time.

The HR diagram can also be used to study the composition of stars. Since a star's temperature and luminosity are related to its internal structure, astronomers can use the HR diagram to infer a star's composition based on its position on the diagram.

In addition, the HR diagram has practical applications in astronomy. For example, it can be used to estimate the distance to a star or group of stars. Since a star's luminosity is related to its distance, astronomers can use the HR diagram to determine the distance to a star based on its position on the diagram.

In conclusion, the Hertzsprung-Russell diagram is a powerful tool for understanding the properties of stars. By plotting a star's luminosity versus its temperature, astronomers can learn a great deal about a star's properties, behavior, and evolution. The HR diagram is an essential tool for studying stars and their role in the universe.

Chapter 21: Binary Stars - Two Stars that Orbit Each Other, and What We Can Learn From Them

The majority of stars in the universe exist as part of a binary or multiple star system, meaning that they have at least one companion star that they orbit around. Binary stars provide astronomers with a wealth of information about the properties and behaviors of stars that cannot be obtained from single stars alone. In this chapter, we will explore the different types of binary star systems, how they are formed, and what we can learn from them.

Types of Binary Star Systems

Binary star systems can be classified based on the separation between the stars, their relative brightness, and their orbital characteristics. The four main types of binary star systems are:

1. **Visual Binaries:** Visual binaries are the easiest type of binary star system to observe, as the two stars can be distinguished visually through a telescope. The stars in a visual binary are relatively far apart, and their orbits can be easily determined.
2. **Spectroscopic Binaries:** In a spectroscopic binary system, the stars are too close together to be visually resolved, but their presence can be detected by observing the variations in their spectral lines as they orbit each other. By analyzing

these spectral lines, astronomers can determine the masses and orbital characteristics of the stars.

3. **Eclipsing Binaries:** Eclipsing binaries are binary star systems in which the stars periodically eclipse each other as seen from Earth, causing a dip in the system's overall brightness. By measuring the time and depth of these eclipses, astronomers can determine the size, mass, and orbital characteristics of the stars.

4. **Astrometric Binaries:** Astrometric binaries are binary star systems in which the motion of one star can be detected by observing its effect on the position of the other star. This effect is typically too small to observe directly, but can be detected through precise measurements of the stars' positions over time.

Formation of Binary Star Systems

Binary star systems can form in a variety of ways, but the most common mechanism is through the fragmentation of a molecular cloud during the process of star formation. As the cloud collapses under its own gravity, it begins to spin, flattening into a disk. The disk eventually breaks up into clumps, each of which can form a star or a binary star system.

In some cases, binary star systems can also form through close encounters between stars. If two stars pass close enough to each other, they can become gravitationally bound and begin to orbit each other.

What We Can Learn from Binary Star Systems

Binary star systems provide astronomers with a wealth of information about the properties and behaviors of stars. By measuring the orbits of binary stars, astronomers can determine the masses of the stars, as well as their sizes and densities. They can also study the interactions between the stars, including the transfer of matter from one star to another and the effects of tidal forces.

Binary star systems are also important for our understanding of stellar evolution. By observing binary stars with different masses and ages, astronomers can study the effects of mass transfer and other processes on the evolution of stars. This allows them to test and refine models of stellar evolution, which in turn helps us to better understand the properties of stars and the history of the universe.

In conclusion, binary star systems are an important area of study in astronomy, providing us with valuable information about the properties and behaviors of stars that cannot be obtained from single stars alone. By continuing to observe and study these systems, we can deepen our understanding of the universe and the processes that shape it.

Chapter 22: Star Clusters - Groups of Stars That Share a Common Origin

Star clusters are fascinating objects that provide important information about the formation and evolution of stars. They are groups of stars that are gravitationally bound and share a common origin. There are two main types of star clusters: open clusters and globular clusters.

Open clusters are young clusters that are found in the disk of our Milky Way galaxy. They are typically made up of a few hundred stars that formed from the same cloud of gas and dust. Open clusters are often visible to the naked eye and can be observed with a small telescope. Some famous examples of open clusters include the Pleiades and the Beehive cluster.

Globular clusters, on the other hand, are much older and contain tens of thousands to millions of stars. They are found in the halo of our galaxy and are tightly packed together due to their strong gravitational attraction. Globular clusters are much more difficult to observe than open clusters, as they are located far away from Earth and require a large telescope to study in detail. Some famous examples of globular clusters include the Hercules Cluster and the Omega Centauri Cluster.

Studying star clusters provides important information about the formation and evolution of stars. For example, by observing the stars in a cluster, astronomers can determine the cluster's age, as well as the ages of individual stars within the cluster. This is because all the stars in a cluster formed from the same cloud of gas and dust at roughly the same time. Therefore, the properties of the stars in the cluster can tell us about the conditions present in the cloud when the stars formed.

In addition to their age, star clusters can also tell us about the evolution of stars. This is because as stars age, they change in size, temperature, and brightness. By studying the properties of the stars in a cluster, astronomers can observe how these properties change over time, providing insight into the various stages of a star's life.

Star clusters are also important for studying the dynamics of our galaxy. By observing the motion of stars in a cluster, astronomers can determine the cluster's position and velocity relative to the Milky Way. This information can then be used to study the overall structure and dynamics of our galaxy.

In recent years, new techniques and instruments have allowed astronomers to study star clusters in even greater detail. For example, the Hubble Space Telescope has provided high-resolution images of distant clusters, allowing astronomers to study individual stars within the clusters. Additionally, advanced spectroscopic techniques have allowed astronomers to study the chemical compositions of stars in clusters, providing important clues about the conditions present when the stars formed.

In summary, star clusters are fascinating objects that provide important information about the formation and evolution of stars, as well as the dynamics of our galaxy. By studying these objects, astronomers can gain insight into the processes that shape our universe, and deepen our understanding of the cosmos.

Chapter 23: Nebulae - The clouds of gas and dust that give birth to stars.

Nebulae are among the most beautiful and fascinating objects in the night sky. These clouds of gas and dust are the birthplace of stars, and they provide important insights into the early stages of star formation. In this chapter, we'll explore the different types of nebulae and what they can tell us about the universe.

1. What are nebulae?

Nebulae are clouds of gas and dust in space. They are primarily composed of hydrogen gas and dust grains made of carbon, silicon, and other elements. Nebulae can range in size from just a few light-years across to hundreds of light-years across.

2. Types of Nebulae

There are three main types of nebulae: emission nebulae, reflection nebulae, and dark nebulae.

- **Emission Nebulae:** These nebulae emit light due to the ionization of gas within them. This ionization is usually caused by nearby hot stars or supernovae. The most famous emission nebula is the Orion Nebula, which is visible to the naked eye and is located in the constellation of Orion.
- **Reflection Nebulae:** These nebulae do not emit light but instead reflect the light of nearby stars. Reflection nebulae

are usually blue in color due to the scattering of blue light by the dust in the nebula. The famous Pleiades star cluster is surrounded by a reflection nebula.

- **Dark Nebulae:** These nebulae do not emit or reflect light and are instead visible because they block the light of stars behind them. The most famous dark nebula is the Horsehead Nebula, which is located in the constellation of Orion.

3. Star Formation in Nebulae

Nebulae are the birthplace of stars. The gas and dust within a nebula can collapse under its own gravity, creating a dense core where a star can form. As the core collapses, it heats up, and nuclear fusion begins. The star then begins to shine and can eventually clear out the surrounding nebula.

4. Observing Nebulae

Nebulae can be observed using telescopes, both ground-based and space-based. Many nebulae are visible to the naked eye, such as the Orion Nebula and the Lagoon Nebula. Others require a telescope to observe, such as the Crab Nebula and the Tarantula Nebula.

5. Importance of Nebulae

Nebulae are not only beautiful objects to observe but also provide important insights into the early stages of star formation. They can also be used to study the properties of interstellar gas and dust and the effects of ionizing radiation on these materials. Additionally, nebulae are a key component of the cosmic web, which is the large-scale structure of the universe.

In conclusion, nebulae are fascinating and beautiful objects in the night sky that provide important insights into the formation of stars and the properties of interstellar gas and dust. Observing these objects can be a rewarding experience for amateur and professional astronomers alike.

Chapter 24: Supernovae - The explosive deaths of massive stars and their aftermath.

Supernovae are some of the most violent and energetic events in the universe. They occur when massive stars reach the end of their lives and run out of fuel for nuclear fusion in their cores. Without the outward pressure generated by nuclear fusion, the star's gravity causes it to collapse in on itself, creating a shockwave that can cause the star to explode.

There are two main types of supernovae: Type I and Type II. Type I supernovae occur in binary star systems, where a white dwarf star steals material from its companion star, eventually causing the white dwarf to reach a critical mass and explode. Type II supernovae, on the other hand, occur in massive stars that have exhausted all their nuclear fuel and have core densities high enough to allow for the fusion of heavier elements. When this core fusion stops, the core collapses and the outer layers of the star are ejected in a spectacular explosion.

Supernovae are important because they are responsible for creating and dispersing heavy elements like iron, nickel, and gold into the universe. These elements are formed in the extreme conditions of a supernova explosion and are then scattered across space by the explosion's shockwave. This process of creating and dispersing heavy elements is known as nucleosynthesis and is critical to the formation of rocky planets like Earth.

One of the most famous supernovae in history is SN 1054, which was observed by Chinese and Japanese astronomers in 1054 CE. The explosion created a bright burst of light that was visible even during the day and remained visible in the night sky for several weeks. The remnants of SN 1054 are still visible today as the Crab Nebula, a cloud of gas and dust that was left behind after the explosion.

In addition to creating heavy elements, supernovae also play an important role in the evolution of galaxies. The energy and shockwaves from a supernova explosion can trigger the formation of new stars and help to disperse gas and dust throughout a galaxy. This process of galaxy evolution is known as feedback and is critical to understanding how galaxies form and change over time.

In recent years, astronomers have been studying supernovae to try and better understand their properties and the processes that lead to their explosions. By observing the light emitted during a supernova explosion, astronomers can learn about the composition of the star that exploded and the conditions that led to the explosion. This information can then be used to refine our understanding of how stars evolve and what happens when they die.

Overall, supernovae are some of the most fascinating and important events in the universe. From creating heavy elements to shaping the evolution of galaxies, these explosive events have a profound impact on the world around us. By studying supernovae, we can learn more about the history of the universe and the processes that have shaped it into what we see today.

Chapter 25: Black Holes - The Mysterious Objects with Gravity so Strong that Nothing can Escape

Black holes are one of the most intriguing and mysterious objects in the universe. They are so massive and dense that they create a gravitational pull so strong that nothing, not even light, can escape their grasp. In this chapter, we will explore what black holes are, how they are formed, and their unique properties that have fascinated scientists for decades.

What are Black Holes?

A black hole is a region in space where gravity is so strong that nothing can escape its pull, not even light. Black holes are formed when massive stars run out of fuel and collapse under their own weight, creating a singularity – a point in space where the laws of physics as we know them break down. This singularity is surrounded by an event horizon, a boundary beyond which anything that enters is trapped forever.

Types of Black Holes

There are three main types of black holes: stellar, intermediate, and supermassive. Stellar black holes are the most common type and are formed from the collapse of a single massive star. Intermediate black holes have a mass between 100 and 100,000 times that of the sun and are thought to be formed from the merging of several smaller black holes. Supermassive black holes have a mass of millions or even billions of times that of the sun and are found at the center of most galaxies.

Properties of Black Holes

Black holes are characterized by their mass, spin, and charge. Mass is the most fundamental property of a black hole and determines its gravitational pull. Spin is a measure of how fast a black hole is rotating, and charge is a measure of the electric charge of the black hole.

Black holes also have unique effects on the space around them. They bend the fabric of spacetime, creating a phenomenon known as gravitational lensing. They can also emit radiation, known as Hawking radiation, due to quantum effects near the event horizon.

Detecting Black Holes

Black holes themselves cannot be directly observed since nothing can escape their gravity. However, their effects on nearby matter can be observed. For example, if a black hole is in a binary star system, the black hole's gravity can cause its companion star to emit X-rays. Scientists can also detect the gravitational waves produced when two black holes merge.

Conclusion

Black holes are some of the most mysterious and fascinating objects in the universe. They are created by the collapse of massive stars, and their gravity is so strong that nothing can escape their grasp. Black holes have unique properties that make them fascinating to study, such as their effects on spacetime and their ability to emit radiation. While we have learned a great deal about black holes, there is still much to discover, and they remain one of the most exciting areas of research in astrophysics.

Chapter 26: Neutron Stars and Pulsars

Neutron stars and pulsars are fascinating astronomical objects that are formed from the remnants of massive stars. These objects are incredibly dense and have strong magnetic fields, making them some of the most extreme environments in the universe. In this chapter, we will explore the properties and behavior of neutron stars and pulsars, and the ways in which they have contributed to our understanding of the universe.

Formation of Neutron Stars

When a massive star reaches the end of its life, it can no longer sustain nuclear fusion in its core, and the core collapses under the force of gravity. The outer layers of the star are blown away in a supernova explosion, leaving behind a compact remnant. If the remnant has a mass of about 1.4 times that of the Sun, it will become a neutron star.

Neutron stars are incredibly dense, with a mass of about 1.4 times that of the Sun compressed into a sphere just 20 kilometers in diameter. This means that neutron stars have a density of about 10^{17} kg/m^3, which is comparable to the density of atomic nuclei.

Properties of Neutron Stars

The extreme density of neutron stars gives rise to a number of interesting properties. For example, neutron stars have incredibly strong magnetic fields, which can be millions of times stronger than Earth's magnetic field. These magnetic fields give rise to powerful emissions of radiation, including X-rays and gamma rays.

Neutron stars also have a very high surface temperature, despite their small size. This is due to the fact that the remnant of the supernova explosion heats up as it collapses, and the resulting neutron star can have a surface temperature of over a million degrees Celsius.

Pulsars

Pulsars are a type of neutron star that emit regular pulses of radiation. These pulses are produced by the rotation of the neutron star, which acts like a lighthouse, beaming radiation into space as it spins. Pulsars were first discovered in 1967 by Jocelyn Bell Burnell and Antony Hewish, who observed a regular pattern of radio signals coming from a source in the constellation Vulpecula.

The discovery of pulsars was a major breakthrough in astronomy, as it provided strong evidence for the existence of neutron stars. Pulsars have also been used to test a number of predictions of general relativity, including the behavior of gravitational waves.

Millisecond Pulsars

Some pulsars rotate at incredible speeds, spinning hundreds of times per second. These are known as millisecond pulsars, and they are thought to be formed through a process called accretion. In an accreting binary system, matter from a companion star falls onto the neutron star, causing it to spin faster and faster.

Millisecond pulsars are important for a number of reasons. For example, they can be used as incredibly precise clocks, with accuracy on the order of microseconds. This makes them useful for studying the behavior of gravity and for testing theories of relativity.

Conclusion

Neutron stars and pulsars are some of the most extreme objects in the universe, and they have provided astronomers with a wealth of information about the properties of matter under extreme conditions. These objects have also played a crucial role in the development of our understanding of general relativity and the behavior of gravity. As we continue to study neutron stars and pulsars, we are sure to uncover even more fascinating insights into the workings of the universe.

Chapter 27: White Dwarfs - The fate of low-mass stars like our Sun.

When stars like our Sun reach the end of their lives, they do not explode in a spectacular supernova. Instead, they undergo a much gentler process that eventually leads to the formation of a white dwarf.

As the fuel in the star's core is depleted, it can no longer support the weight of the overlying layers. The core begins to contract under the force of gravity, and the outer layers expand outward. This causes the star to become a red giant, much larger and cooler than it was before.

Eventually, the outer layers of the red giant will drift away into space, leaving behind a small, dense core. This core is now a white dwarf, a hot and incredibly dense object that is about the size of Earth but contains the mass of a star.

White dwarfs are incredibly hot, with temperatures that can reach up to 100,000 Kelvin. They are also incredibly dense, with a mass that is typically around 0.6 times that of the Sun but compressed into a space the size of a planet. This means that a teaspoon of material from a white dwarf would weigh about 5 tons!

One of the most interesting things about white dwarfs is their evolution. They don't produce energy through nuclear fusion like regular stars do, but instead slowly cool down over time. As they cool, they change color, starting out as blue-white and eventually becoming red. This means that by looking at the color of a white dwarf, astronomers can determine its age.

Another interesting aspect of white dwarfs is their relationship to planetary systems. As a star like our Sun evolves into a red giant, it will expand and may engulf any planets that are orbiting it. However, if the planets are far enough away, they can survive the red giant phase and remain in orbit around the white dwarf. This means that white dwarfs can have planets, and in fact, some of the first exoplanets ever discovered were found orbiting white dwarfs.

In addition, white dwarfs can also undergo a type of explosion known as a nova. This happens when a white dwarf is in a binary system with another star and begins to siphon material off of its companion. This material can build up on the surface of the white dwarf, eventually causing a runaway nuclear reaction that leads to a sudden, bright outburst of light.

Overall, white dwarfs are fascinating objects that have much to teach us about the end stages of a star's life and the evolution of planetary systems.

Chapter 28: Galaxies - The different types of galaxies and their properties

Introduction: The study of galaxies is one of the most fascinating aspects of astronomy. A galaxy is a vast collection of stars, gas, and dust that is held together by gravity. There are billions of galaxies in the observable universe, ranging in size from small dwarf galaxies to massive elliptical galaxies that contain trillions of stars. In this chapter, we will explore the different types of galaxies and their properties.

Types of galaxies: Galaxies can be broadly classified into three main types: spiral, elliptical, and irregular. Let's take a closer look at each type.

1. **Spiral galaxies:** Spiral galaxies are characterized by a central bulge surrounded by spiral arms that contain young, hot stars and interstellar gas and dust. Our own Milky Way galaxy is a spiral galaxy. Spiral galaxies can be further classified into two subtypes: barred and unbarred. Barred spiral galaxies have a central bar structure that runs through the bulge and connects to the spiral arms.

2. **Elliptical galaxies:** Elliptical galaxies are shaped like a flattened sphere or an egg. They have a smooth, featureless appearance and contain mostly old stars. Elliptical galaxies can range in size from small dwarf galaxies to massive giants that can contain trillions of stars.

3. **Irregular galaxies:** Irregular galaxies have a chaotic, irregular shape and no distinct bulge or arms. They are often small and contain young stars and interstellar gas and dust. Irregular galaxies can be further classified into two subtypes: dwarf irregular and giant irregular.

Properties of galaxies: Galaxies have a variety of properties that astronomers use to classify and study them. Some of the important properties include:

1. **Size:** Galaxies can range in size from small dwarf galaxies to massive giants that can contain trillions of stars.
2. **Luminosity:** The total amount of energy emitted by a galaxy is known as its luminosity. Galaxies can have a wide range of luminosities, depending on the number and types of stars they contain.
3. **Color:** The color of a galaxy can provide information about its age and the types of stars it contains. Young, hot stars emit blue light, while older, cooler stars emit redder light.
4. **Redshift:** The redshift of a galaxy is a measure of how much its light has been shifted towards the red end of the spectrum due to its motion away from us. This can provide information about the distance and velocity of the galaxy.

Conclusion:

In conclusion, galaxies are fascinating objects that come in a variety of shapes and sizes. Understanding the different types of galaxies and their properties can provide important insights into the evolution and structure of the universe. With new technologies and instruments being developed all the time, we can expect to learn even more about these amazing objects in the years to come.

Chapter 29: The Milky Way

Our home in the universe is the Milky Way galaxy, a vast, swirling collection of stars, gas, and dust that stretches across hundreds of thousands of light-years. In this chapter, we will explore the structure, properties, and history of our galaxy.

The Milky Way is a barred spiral galaxy, which means that it has a central bar-shaped region surrounded by spiral arms. It is estimated to contain between 100 and 400 billion stars, although this number is difficult to determine accurately. The stars in the Milky Way are organized into different structures, including the central bulge, the disk, and the halo.

The central bulge is a dense, spherical region at the center of the Milky Way that contains mostly old stars. It is thought to have formed early in the galaxy's history through the collapse of a gas cloud. The disk of the Milky Way is a flattened, rotating structure that contains most of the galaxy's stars, as well as gas and dust. The spiral arms are regions of increased star formation that wind around the disk. The halo is a spherical region surrounding the disk that contains mostly old stars and globular clusters.

The Milky Way is also home to several other interesting structures, including the galactic bar and the galactic center. The bar is an elongated feature that extends from the central bulge through the disk, and is thought to be responsible for the formation and maintenance of the spiral arms. The galactic center is a region at the very center of the Milky Way that contains a supermassive black hole, as well as several other interesting objects, such as the Sagittarius A* radio source and a cluster of stars known as the S-stars.

The Milky Way has a complex history that has been shaped by a variety of factors, including the formation of stars and the interactions between different structures within the galaxy. It is believed to have formed about 13.6 billion years ago through the collapse of a large gas cloud. Over time, stars and other structures have formed within the galaxy through a variety of processes, including the accretion of gas and dust, the merger of smaller galaxies, and the formation of globular clusters.

The study of the Milky Way is essential to our understanding of the universe as a whole. By studying the properties and behavior of stars, gas, and other structures within the galaxy, astronomers can gain insight into the physical processes that shape galaxies and the universe as a whole. Additionally, the study of the Milky Way can help us better understand our place in the universe, and the factors that have influenced the formation and evolution of life on Earth.

In recent years, advances in technology and observational techniques have allowed astronomers to make significant progress in our understanding of the Milky Way. Large-scale surveys of the galaxy, such as the Sloan Digital Sky Survey and the Gaia mission, have provided detailed maps of the positions and movements of millions of stars within the Milky Way. These surveys have allowed astronomers to study the structure and properties of the galaxy in unprecedented detail, and have provided insights into the formation and evolution of the Milky Way and other galaxies.

In conclusion, the Milky Way is a fascinating and complex object that has captivated human curiosity for centuries. As our understanding of the galaxy and the universe as a whole continues to evolve, we can expect to make even more exciting discoveries about the history and properties of our home in the universe.

Chapter 30: Dark Matter

When astronomers observe the universe, they can detect the presence of matter by the light it emits or absorbs. However, they have found that the amount of visible matter in the universe does not account for the observed gravitational effects on objects. This has led to the hypothesis that there is another type of matter that is invisible to telescopes, called dark matter. In this chapter, we will explore what dark matter is, how it was discovered, and its importance in understanding the universe.

Discovery of Dark Matter

The idea of dark matter was first proposed in the 1930s by Swiss astronomer Fritz Zwicky. He studied the Coma Cluster, a group of galaxies that were moving too fast for the visible matter to hold them together. He suggested that there must be a large amount of unseen matter that was causing the galaxies to stay in the cluster. Zwicky estimated that this dark matter made up about 400 times more mass than the visible matter.

However, it was not until the 1970s that the concept of dark matter gained widespread acceptance in the scientific community. Vera Rubin, an American astronomer, observed the rotation curves of galaxies and found that the visible matter was not enough to explain the observed velocities. The stars in the outer regions of galaxies were moving too fast, and there had to be more mass present that was not visible. Rubin's observations provided strong evidence for the existence of dark matter.

Types of Dark Matter

The nature of dark matter remains a mystery, but there are several theories about what it could be made of. One possibility is that it is composed of weakly interacting massive particles (WIMPs), which would be left over from the Big Bang. These particles would be very difficult to detect, as they would interact very weakly with ordinary matter.

Another theory is that dark matter is made up of massive compact halo objects (MACHOs), such as black holes, brown dwarfs, or planets. However, recent observations have ruled out this possibility as the dominant form of dark matter.

Dark Matter and the Universe

The presence of dark matter has significant implications for our understanding of the universe. One of the most important is its role in the formation of large-scale structures. The gravity of dark matter was the dominant force in the early universe, pulling matter together and forming the first galaxies and clusters of galaxies.

Dark matter is also important in explaining the observed distribution of matter in the universe. The large-scale structures, such as galaxy clusters and filaments, are thought to have formed along the densest regions of dark matter. This has been confirmed by observations of the cosmic microwave background radiation, which shows the distribution of matter in the early universe.

Furthermore, the existence of dark matter affects the fate of the universe. If there is enough dark matter, its gravity will eventually slow down the expansion of the universe and cause it to collapse. If there is not enough, the universe will continue to expand forever.

Detecting Dark Matter

Despite its significance, dark matter has never been directly detected. Scientists are currently working on several experiments to detect WIMPs, which would confirm their existence. One experiment involves detecting the radiation given off when WIMPs collide with atoms in a detector. Other experiments involve detecting the effect of dark matter on the motion of stars in galaxies or the distortion of light from distant galaxies.

Conclusion

Dark matter remains one of the greatest mysteries in modern astronomy. Its presence is inferred from the observed gravitational effects on visible matter, but it has never been directly detected. Nevertheless, the existence of dark matter is critical to our understanding of the universe, from the formation of galaxies to the fate of the cosmos. Scientists continue to search for ways to detect and understand this elusive substance.

Chapter 31: Dark Energy

The discovery of the accelerated expansion of the universe in the late 1990s was one of the most surprising and revolutionary findings in the history of astronomy. It implied the existence of a new and unknown type of energy that was driving this expansion, which has since been named "dark energy." The nature and properties of dark energy are still unknown, and it remains one of the biggest mysteries in modern astrophysics.

Discovery of Dark Energy:

The discovery of dark energy came from observations of distant supernovae by two independent teams, the Supernova Cosmology Project and the High-Z Supernova Search Team. These teams were trying to measure the deceleration of the expansion of the universe using the brightness of Type Ia supernovae, which are known to have a consistent intrinsic brightness. However, to their surprise, they found that these supernovae were fainter than expected, indicating that the universe was not decelerating, but instead, it was accelerating in its expansion.

The implications of this discovery were profound, as it implied the existence of a new and unknown form of energy that was driving this acceleration. The name "dark energy" was coined to describe this unknown energy, and it is believed to make up about 68% of the total energy density of the universe.

Properties of Dark Energy:

The nature of dark energy is still unknown, and it is one of the biggest mysteries in modern astrophysics. However, there are some proposed theories that attempt to explain its properties. The most popular theory is that dark energy is a property of space itself, known as the cosmological constant. This theory proposes that empty space has a non-zero energy density that exerts a repulsive force on matter, which is responsible for the observed accelerated expansion of the universe.

Another theory proposes that dark energy is a new scalar field, similar to the Higgs field, which permeates all of space and causes the acceleration of the universe. However, this theory requires new physics beyond the standard model to explain the properties of this scalar field.

Implications of Dark Energy:

The discovery of dark energy has had significant implications for our understanding of the universe. It has led to a new and revised understanding of the history and fate of the universe. The accelerating expansion implies that the universe will continue to expand indefinitely, and eventually, the galaxies will become so far apart that they will no longer be visible to each other. This scenario, known as the "Big Freeze," is currently the most accepted theory for the fate of the universe.

Additionally, dark energy has implications for the large-scale structure of the universe. It is believed that the distribution of dark matter and dark energy determines the large-scale structure of the universe. Understanding the nature of dark energy is, therefore, crucial to understanding the formation and evolution of galaxies and other structures in the universe.

Conclusion:

Dark energy remains one of the biggest mysteries in modern astrophysics. Its discovery has revolutionized our understanding of the universe, and it has significant implications for the history and fate of the universe, as well as the large-scale structure of the universe. While there are proposed theories to explain its nature, the true properties of dark energy remain unknown, and further observations and studies are needed to unravel this enigma of the universe.

Chapter 32: The Big Bang - The origin of the universe and the evidence supporting this theory.

The Big Bang theory is the prevailing cosmological model for the origin and evolution of the universe. It states that the universe began as a singular, extremely hot and dense point, known as a singularity, around 13.8 billion years ago. The universe has been expanding and cooling ever since, and the current observed universe can be traced back to the Big Bang.

Evidence for the Big Bang theory includes the cosmic microwave background radiation (CMB), the abundance of light elements such as helium and hydrogen, and the large-scale structure of the universe.

The cosmic microwave background radiation is one of the strongest pieces of evidence for the Big Bang theory. It is a faint glow of radiation that fills the entire universe, and is thought to be the remnants of the Big Bang's initial explosion. The CMB was first detected in 1964 by radio astronomers Arno Penzias and Robert Wilson, who were awarded the Nobel Prize in Physics in 1978 for their discovery.

The abundance of light elements, such as helium and hydrogen, is another piece of evidence for the Big Bang theory. These elements were formed during the first few minutes after the Big Bang when the universe was still hot and dense enough to allow for nuclear fusion to occur. The precise ratios of the abundance of these light elements can be accurately predicted based on the conditions present during this time.

The large-scale structure of the universe also provides evidence for the Big Bang theory. The universe is observed to be homogeneous and isotropic on very large scales, meaning that it looks the same in all directions and at all locations. This is consistent with the idea that the universe began as a uniform and homogenous singularity that has since expanded and cooled.

The Big Bang theory has undergone several refinements since its inception, including the theory of cosmic inflation, which suggests that the universe underwent a brief period of exponential expansion shortly after the Big Bang. This theory helps to explain certain observations that are difficult to reconcile with the standard Big Bang theory.

The Big Bang theory has also led to a better understanding of the history and evolution of the universe. It explains the origin of the cosmic microwave background radiation, the formation and evolution of galaxies, and the abundance of heavier elements in the universe. The Big Bang theory is the foundation for modern cosmology and has inspired generations of researchers to explore the origins of the universe.

In conclusion, the Big Bang theory is the prevailing cosmological model for the origin and evolution of the universe. It is supported by a wealth of observational evidence, including the cosmic microwave background radiation, the abundance of light elements, and the large-scale structure of the universe. The Big Bang theory has undergone refinements over the years, but remains the most widely accepted explanation for the origin of the universe.

Chapter 33: Cosmic Microwave Background Radiation

The cosmic microwave background radiation (CMB) is a faint, uniform glow of electromagnetic radiation that permeates the entire universe. It is considered one of the strongest pieces of evidence for the Big Bang theory of the universe's origin. The CMB was discovered accidentally in 1964 by two radio astronomers, Arno Penzias and Robert Wilson, who were using a large radio telescope at Bell Labs in New Jersey to search for faint radio signals from outer space. They found a mysterious noise that seemed to be coming from all directions in the sky, and they couldn't eliminate it no matter what they did.

It wasn't until a few months later that they learned about the work of physicists Robert Dicke and Jim Peebles at Princeton University, who had predicted the existence of the CMB as a remnant of the Big Bang. Penzias and Wilson realized that they had accidentally discovered this radiation, which had been predicted to exist but had never been detected before.

The CMB has a temperature of about 2.73 Kelvin (or -270.42°C) and is almost perfectly uniform in all directions, with tiny fluctuations in temperature that are believed to be caused by quantum fluctuations in the early universe. These fluctuations were imprinted on the CMB when the universe was only about 380,000 years old, and they provide important clues about the universe's structure and history.

The discovery of the CMB revolutionized cosmology and provided strong support for the Big Bang theory, which predicts that the universe began as a hot, dense, and uniform state and has been expanding and cooling ever since. The CMB is often described as the "echo" of the Big Bang, as it allows us to see back in time to when the universe was only a few hundred thousand years old.

One of the key features of the CMB is its blackbody spectrum, which is a curve that describes the distribution of radiation as a function of wavelength. A blackbody spectrum is produced by a hot, dense object that is in thermal equilibrium, and it has a characteristic shape that depends only on the object's temperature. The CMB spectrum is almost a perfect blackbody, with a temperature of about 2.73 Kelvin. This is strong evidence that the CMB was produced by a hot, dense, and uniform state, as predicted by the Big Bang theory.

Another important feature of the CMB is the pattern of temperature fluctuations across the sky. These fluctuations are extremely small, only about one part in 100,000, but they contain a wealth of information about the universe's structure and history. The pattern of fluctuations is believed to have been imprinted on the CMB by quantum fluctuations in the early universe, which were amplified by the universe's expansion and eventually grew into the large-scale structures we see today, such as galaxies and galaxy clusters.

Cosmologists study the CMB in great detail to learn about the universe's structure and history. By analyzing the pattern of temperature fluctuations, they can determine important parameters such as the universe's age, its expansion rate, and the amount of dark matter and dark energy it contains. They can also test different cosmological models and theories by comparing their predictions to the observed CMB data.

In recent years, several experiments have been designed and launched to study the CMB in even greater detail. These experiments include the Wilkinson Microwave Anisotropy Probe (WMAP) and the European Space Agency's Planck satellite, which have provided some of the most precise measurements of the CMB to date. These experiments have confirmed many of the predictions of the Big Bang theory and

In conclusion, the cosmic microwave background radiation is a crucial piece of evidence supporting the Big Bang theory. It provides insight into the early universe and the conditions that existed shortly after the Big Bang. The CMBR also confirms the universe's overall isotropy and homogeneity, and the slight variations in temperature allow us to study the universe's structure and evolution. With advances in technology and new observational data, scientists continue to study the CMBR, and it remains an essential tool for understanding the universe's origins and evolution.have led to new insights into the nature of the universe.

Chapter 34: The Cosmic Web - The large-scale structure of the universe.

The universe is not just a random scattering of stars, galaxies, and other structures. Instead, these objects are organized into a vast and intricate pattern known as the cosmic web. The cosmic web is a web-like structure of galaxies and galaxy clusters connected by vast filaments of dark matter and gas, spanning billions of light-years.

The cosmic web is a fundamental feature of the universe, and it provides us with crucial insights into the history and evolution of the cosmos. It is thought to have formed through the process of gravitational collapse, in which dark matter and gas clump together under the influence of their own gravity.

At the largest scales, the cosmic web appears to be made up of clusters and superclusters of galaxies, connected by vast filaments of gas and dark matter. These filaments can stretch for hundreds of millions of light-years and contain enormous quantities of gas and dust. They are thought to be the sites of ongoing star formation and to be responsible for the distribution of heavy elements in the universe.

The cosmic web is not just a static structure. It is constantly evolving, with galaxies and galaxy clusters moving around and interacting with each other under the influence of gravity. These interactions can cause galaxies to merge and form larger structures, such as superclusters of galaxies.

Observing the cosmic web is challenging, as its structure is mostly invisible to our eyes. However, astronomers have developed sophisticated techniques for mapping the distribution of matter in the universe. One such technique is gravitational lensing, which involves observing the bending of light by massive objects, such as galaxy clusters. By studying the distortions in the light, astronomers can infer the location and distribution of dark matter and gas in the cosmic web.

Studying the cosmic web is essential to our understanding of the universe. By mapping out its structure and evolution, we can gain insights into the processes that shaped the cosmos, from its earliest moments to the present day. It is also important for understanding the distribution of matter and energy in the universe, and for tracing the evolution of galaxies and galaxy clusters over billions of years.

Chapter 35: Gravitational Waves - Ripples in space-time caused by violent cosmic events

Gravitational waves are ripples in the fabric of space-time caused by the acceleration of massive objects. These waves were predicted by Einstein's theory of general relativity in 1915, but it wasn't until 2015 that they were first directly detected by the Laser Interferometer Gravitational-Wave Observatory (LIGO) in the United States. Since then, many more detections have been made by LIGO and its European counterpart, Virgo.

Gravitational waves are produced by some of the most violent events in the universe, such as the collision of black holes or the explosion of supernovae. These events cause ripples in the fabric of space-time that spread out across the universe at the speed of light. When they reach Earth, they cause incredibly tiny fluctuations in the distances between objects that can be detected by instruments like LIGO and Virgo.

The detection of gravitational waves has opened up a new way of studying the universe. Before the discovery of gravitational waves, astronomers could only observe the universe using electromagnetic radiation, such as light and radio waves. Gravitational waves, on the other hand, allow astronomers to observe the universe using a completely different medium. By detecting the ripples in space-time caused by violent cosmic events, astronomers can study the properties of the objects that produced them, such as their mass, spin, and distance from Earth.

One of the most exciting discoveries made using gravitational waves is the detection of black hole mergers. Black holes are some of the most mysterious objects in the universe, and until the detection of gravitational waves, there was no direct evidence that they existed. However, the detection of gravitational waves from black hole mergers has provided strong evidence for the existence of these enigmatic objects.

Gravitational waves have also been used to study neutron stars, which are the ultra-dense remnants of supernovae. In 2017, LIGO and Virgo detected gravitational waves from the collision of two neutron stars. This event also produced a burst of gamma rays, which was detected by several space-based telescopes. This marked the first time that both gravitational waves and electromagnetic radiation had been detected from the same cosmic event. The observation of this event has provided valuable insights into the properties of neutron stars and the nature of the universe.

In addition to black hole mergers and neutron star collisions, gravitational waves have also been used to study other cosmic phenomena, such as the early universe, pulsars, and even the nature of gravity itself.

The study of gravitational waves is still in its infancy, and there is much to be learned about these ripples in space-time.

However, the detection of gravitational waves has already revolutionized our understanding of the universe and opened up new avenues for exploration. With ongoing improvements in technology and more powerful detectors, astronomers are sure to make many more exciting discoveries using gravitational waves in the years to come.

Chapter 36: Exoplanets - Planets that orbit other stars, and the search for life beyond Earth.

For centuries, humans have wondered whether there are other planets in the universe like our own, and whether there might be life on them. In recent decades, thanks to advances in technology, we have been able to discover planets beyond our solar system, known as exoplanets. These discoveries have revolutionized our understanding of the universe and opened up a whole new field of research.

The first confirmed exoplanet was discovered in 1992, orbiting a pulsar known as PSR B1257+12. Since then, the number of confirmed exoplanets has grown exponentially, with thousands of planets now known to orbit other stars. These exoplanets come in a wide range of sizes, compositions, and orbits, and have led to new insights into planetary formation and evolution.

The most common method of detecting exoplanets is known as the transit method. This involves observing a star and looking for a periodic dip in its brightness, which can be caused by a planet passing in front of it. Another method is the radial velocity method, which involves measuring the slight wobble of a star caused by the gravitational pull of an orbiting planet.

Exoplanets have been found in a wide range of environments, from hot, Jupiter-like gas giants orbiting close to their stars to icy, Earth-sized planets orbiting far away. Many of these planets are part of multi-planet systems, with some systems having up to seven or eight planets orbiting a single star.

One of the most exciting aspects of exoplanet research is the search for habitable planets – planets that are similar to Earth and may have the right conditions for life to exist. In 2015, NASA's Kepler space telescope discovered Kepler-452b, a planet about 1.5 times the size of Earth that orbits a sun-like star at a similar distance to Earth's orbit around the Sun. This led to speculation that Kepler-452b may be a potentially habitable planet, although further research is needed to confirm this.

In addition to searching for potentially habitable planets, exoplanet research has also led to the discovery of new types of planets, such as hot Jupiters, mini-Neptunes, and super-Earths. These planets have expanded our understanding of planetary formation and evolution, and have challenged existing theories about how planets form and migrate within their star systems.

One of the most exciting developments in exoplanet research is the search for signs of life on other planets. This involves looking for biosignatures – molecules or atmospheric conditions that could indicate the presence of life. In 2017, NASA's Hubble Space Telescope detected the presence of water vapor in the atmosphere of a potentially habitable exoplanet known as TRAPPIST-1e. This was a major breakthrough in the search for habitable exoplanets, although further research is needed to confirm the presence of life.

In addition to ground-based telescopes and space-based observatories like Kepler and Hubble, new and upcoming missions will further expand our understanding of exoplanets. These include the James Webb Space Telescope, which is set to launch in 2021, and the European Space Agency's PLATO mission, which will search for Earth-sized exoplanets in the habitable zones of other stars.

The discovery of exoplanets has revolutionized our understanding of the universe and our place within it. These planets offer a glimpse into the incredible diversity of planetary systems beyond our own, and have opened up exciting new avenues of research in the search for life beyond Earth. As technology continues to advance, we can only imagine what other discoveries await us in the fascinating world of exoplanet research.

Chapter 37: Habitable Zones - The regions around a star where liquid water could exist on a planet's surface

One of the main criteria for searching for habitable exoplanets is the presence of liquid water on their surface. Liquid water is considered essential for life as we know it, and therefore, the habitable zone (HZ) of a star is a critical factor in determining whether a planet could potentially support life. In this chapter, we will explore what the habitable zone is, how it is defined, and what factors can affect its size and location.

What is the Habitable Zone?

The habitable zone is defined as the region around a star where a planet can have surface temperatures that allow liquid water to exist. The presence of liquid water is essential because it is a key ingredient for life as we know it. A planet that is too close to its star will be too hot, and water will evaporate, leaving the planet dry. Conversely, a planet that is too far from its star will be too cold, and water will freeze, leaving the planet covered in ice.

The habitable zone is sometimes referred to as the "Goldilocks Zone" because it is not too hot and not too cold, but just right, like the porridge that Goldilocks ate in the famous children's story. In other words, the habitable zone is the region where conditions are just right for life to exist.

How is the Habitable Zone Defined?

The habitable zone is defined based on the amount of energy a planet receives from its star. This energy is determined by the star's luminosity and the distance between the planet and the star. If a planet is too close to its star, it will receive too much energy, and its surface temperature will be too high for liquid water to exist. If a planet is too far from its star, it will receive too little energy, and its surface temperature will be too low for liquid water to exist.

The habitable zone is usually defined as the range of distances from a star where a planet could have surface temperatures between 0°C and 100°C (32°F to 212°F), which is the range where liquid water can exist. However, the size of the habitable zone can vary depending on the properties of the star and the planet.

Factors that Affect the Size and Location of the Habitable Zone

Several factors can affect the size and location of the habitable zone. The most important factor is the star's luminosity. A star's luminosity is the total amount of energy it emits per unit time, and it determines how much energy a planet receives from its star. The habitable zone of a more massive and luminous star is further away than that of a less massive and less luminous star.

Another factor that can affect the size and location of the habitable zone is the planet's atmosphere. A planet with a thick atmosphere can trap more heat and have a wider habitable zone than a planet with a thin atmosphere. However, a thick atmosphere can also lead to a runaway greenhouse effect, where the planet's temperature increases to a point where all water evaporates and the planet becomes uninhabitable.

The planet's albedo, or the amount of light it reflects, can also affect the habitable zone. A planet with a high albedo reflects more sunlight and can have a wider habitable zone than a planet with a low albedo. Finally, the presence of a moon or other large body can affect the habitable zone. A moon can stabilize a planet's orbit and prevent it from experiencing extreme temperature fluctuations that would make it uninhabitable.

Conclusion

In conclusion, the concept of a habitable zone has revolutionized our search for life beyond Earth. It allows us to narrow down the search to the regions around a star where the conditions are just right for liquid water to exist on a planet's surface. This is a crucial factor in determining whether a planet is capable of supporting life as we know it.

The study of habitable zones has also led to a better understanding of the conditions necessary for life to thrive, and has helped scientists refine their methods for detecting and characterizing exoplanets. With the continued search for exoplanets and the development of new technologies, it is possible that we will soon discover a planet in a habitable zone that could harbor life.

However, it is important to remember that the habitable zone is just one of many factors that determine a planet's ability to support life. The planet's atmosphere, magnetic field, and other factors also play a crucial role in creating and maintaining habitable conditions. As we continue to study exoplanets and search for signs of life, we must consider all of these factors and remain open to the possibility of life existing in environments that are different from those we are familiar with on Earth.

Chapter 38: SETI - The search for extraterrestrial intelligence and how we are looking for it.

The possibility of finding intelligent life beyond Earth has fascinated scientists and the general public alike for decades. The Search for Extraterrestrial Intelligence, or SETI, is a field of research dedicated to detecting signals from extraterrestrial civilizations. SETI researchers use a range of methods and technologies to search for signals from other worlds, including radio telescopes, optical telescopes, and sophisticated computer algorithms.

The idea of searching for extraterrestrial life is not new. In fact, scientists have been exploring this possibility for centuries. In the early 17th century, Johannes Kepler suggested that other planets in the universe might be inhabited. In the 19th century, scientists began to speculate about the possibility of life on Mars. In the 20th century, the advent of radio astronomy provided a new tool for searching for signals from other worlds, leading to the birth of the SETI field.

One of the primary methods used in SETI is radio astronomy. This involves searching for radio signals that could be indicative of an extraterrestrial civilization. Radio signals from space are detected using large radio telescopes, which are sensitive to very weak signals from distant objects. Scientists search for signals that are not naturally occurring and cannot be explained by known phenomena.

Another method used in SETI is optical astronomy. This involves searching for signals that could be seen with optical telescopes, such as laser signals or optical flashes. Optical SETI is still in its early stages, and no conclusive evidence of an extraterrestrial signal has been detected using this method.

The search for extraterrestrial intelligence is not limited to the use of telescopes. Scientists also use computer algorithms to analyze vast amounts of data, searching for patterns or anomalies that could be indicative of a signal from an extraterrestrial civilization. This approach is known as signal processing or data mining.

One of the challenges in SETI is distinguishing signals from extraterrestrial civilizations from natural phenomena and human-made signals. To address this challenge, SETI researchers use a variety of techniques to filter out signals that are not of interest, such as signals from Earth-based radio sources or satellites.

Despite decades of searching, no conclusive evidence of an extraterrestrial civilization has been found. However, SETI researchers remain optimistic, and the field continues to evolve with advances in technology and new discoveries. The possibility of finding life beyond Earth remains a tantalizing and exciting prospect.

In addition to the search for extraterrestrial signals, the search for extraterrestrial life also includes the search for habitable exoplanets. The discovery of thousands of exoplanets in recent years has provided new opportunities for SETI researchers to identify targets for their search. The concept of a "habitable zone," or the region around a star where liquid water could exist on a planet's surface, has become a key focus in the search for life beyond Earth.

In conclusion, the search for extraterrestrial intelligence is a fascinating and ongoing field of research. Although we have not yet found conclusive evidence of an extraterrestrial civilization, advances in technology and new discoveries provide reason for continued optimism. The search for life beyond Earth and the possibility of contact with an extraterrestrial civilization remain one of the greatest scientific mysteries of our time.

Chapter 39: Life in the Universe - The factors that make life possible and the potential for life elsewhere.

One of the most fundamental questions humans have asked is whether or not we are alone in the universe. The search for life beyond Earth has captured the imaginations of scientists and the public alike for centuries. In recent years, advances in technology and our understanding of the conditions necessary for life have greatly expanded our search for extraterrestrial life. This chapter will explore the factors that make life possible and the potential for life elsewhere in the universe.

Life on Earth

Before we can discuss the potential for life elsewhere in the universe, we must first understand the conditions necessary for life on Earth. Earth is the only planet in our solar system known to harbor life, and it is a complex and diverse ecosystem. Life on Earth is based on carbon, which is a versatile element that can form a wide range of chemical bonds. Carbon-based molecules are the foundation of all life on Earth, including DNA and proteins.

In addition to carbon, life on Earth also requires a source of energy. The sun provides energy for photosynthesis, the process by which plants convert sunlight into energy. Some organisms, such as bacteria, use chemosynthesis to obtain energy from chemicals in their environment. Water is another crucial ingredient for life on Earth. Water is a solvent that can dissolve many different types of molecules, making it an ideal environment for the chemical reactions necessary for life.

The Potential for Life Elsewhere

Given the conditions necessary for life on Earth, scientists have been searching for similar conditions elsewhere in the universe. The search for life beyond Earth is based on the principle of astrobiology, which is the study of the origin, evolution, distribution, and future of life in the universe. Astrobiologists are particularly interested in finding exoplanets, which are planets that orbit stars other than the sun.

One of the most important factors in the search for life elsewhere in the universe is the concept of the habitable zone. The habitable zone, also known as the Goldilocks zone, is the region around a star where a planet could have the right conditions for liquid water to exist on its surface. Liquid water is considered a key ingredient for life, as it is necessary for many of the chemical reactions that support life.

The habitable zone is not the only factor necessary for life, however. Other factors include the composition of the planet's atmosphere, the presence of organic molecules, and the planet's proximity to its host star. For example, a planet that is too close to its host star may be too hot to support life, while a planet that is too far away may be too cold.

In recent years, advances in technology have allowed scientists to identify thousands of exoplanets. While the search for life beyond Earth is still in its early stages, these discoveries have greatly expanded our understanding of the potential for life elsewhere in the universe. In addition, the discovery of extremophiles, organisms that can survive in extreme environments on Earth, has led scientists to expand their definition of what conditions could support life.

The Search for Extraterrestrial Intelligence

The search for life beyond Earth also includes the search for extraterrestrial intelligence, or SETI. SETI is the scientific search for evidence of intelligent life elsewhere in the universe. This includes the search for radio signals, optical signals, and other forms of communication.

The SETI program has been ongoing since the 1960s, but so far no evidence of extraterrestrial intelligence has been found. However, the search continues, and advances in technology are allowing scientists to search for signals with greater sensitivity and over a wider range of frequencies.

Conclusion

In conclusion, the search for life in the universe is a complex and exciting field that requires a multidisciplinary approach. The study of astrobiology involves a combination of astronomy, biology, chemistry, and geology to understand the conditions that make life possible and to search for signs of life beyond Earth. The discovery of exoplanets, the development of new telescopes, and advances in technology have greatly expanded our ability to search for life in the universe. While the search for life beyond Earth is still ongoing, the potential for finding life elsewhere in the universe is tantalizing. The search for extraterrestrial life not only has the potential to revolutionize our understanding of the universe but also to inspire a new generation of scientists and engineers to continue the search.

Chapter 40: Astrobiology - The study of the origin, evolution, and distribution of life in the universe.

Astrobiology is a multidisciplinary field that seeks to answer the fundamental question of whether life exists beyond Earth. It draws on the fields of biology, chemistry, geology, astronomy, and planetary science to investigate the possibility of life on other planets and moons.

The search for life beyond Earth is motivated by the fact that life on our planet is diverse and resilient, capable of thriving in extreme environments such as deep-sea hydrothermal vents, polar ice caps, and acidic hot springs. Furthermore, life on Earth has existed for over 3.5 billion years, suggesting that the conditions for life may be common in the universe.

One of the key areas of study in astrobiology is the identification of habitable environments beyond Earth. A habitable environment is one that has the necessary conditions to support life as we know it. This includes the presence of liquid water, a source of energy, and a stable environment. The habitable zone around a star, also known as the Goldilocks zone, is the range of distances where a planet could maintain liquid water on its surface.

The search for habitable environments has focused on our own solar system, with the identification of potential habitats on

Mars, Europa, and Enceladus. Mars is of particular interest due to its similarity to Earth in terms of its size, composition, and history. Mars once had a thicker atmosphere and liquid water on its surface, leading to speculation that life may have existed there in the past.

Beyond our solar system, the search for habitable environments involves the detection and characterization of exoplanets. The Kepler space telescope has been instrumental in this regard, identifying thousands of exoplanets and hundreds of potential habitable planets. The James Webb Space Telescope, set to launch in 2021, will enable scientists to study the atmospheres of exoplanets and search for signs of life.

Another area of study in astrobiology is the origins of life on Earth. Scientists believe that life on Earth originated from simple organic compounds that formed in the early oceans and were subsequently concentrated and transformed by geological processes. The conditions on early Earth were harsh, with high temperatures, intense radiation, and frequent asteroid impacts, yet life still managed to emerge.

Understanding the origins of life on Earth provides insights into the conditions necessary for life to arise elsewhere in the universe. It also sheds light on the possibility of panspermia, the hypothesis that life on Earth may have been seeded by microorganisms from space.

Astrobiology also considers the potential for life beyond the traditional forms that we know. The discovery of extremophiles, organisms that thrive in extreme environments, has expanded our understanding of the conditions necessary for life. This has led to speculation about the existence of exotic life forms, such as organisms that use alternative biochemistries or are based on silicon rather than carbon.

In conclusion, astrobiology is a fascinating and rapidly developing field that seeks to answer some of the most profound questions about our place in the universe. While the search for extraterrestrial life remains ongoing, astrobiology has already provided insights into the conditions necessary for life, the origins of life on Earth, and the potential for exotic forms of life. With advances in technology and increased exploration of our solar system and beyond, we are likely to uncover even more surprises in the coming years.

Chapter 41: Colonizing Other Planets - The challenges and opportunities of living on other planets and moons.

As humans continue to explore the universe and search for habitable worlds, the idea of colonizing other planets has become a topic of increasing interest. The potential benefits of establishing colonies on other planets or moons are numerous, including expanding the reach of human civilization, ensuring the survival of our species in the event of a global catastrophe on Earth, and furthering scientific exploration and discovery.

However, the challenges of colonizing other planets are significant, and will require innovative solutions and technologies to overcome. In this chapter, we will explore some of the challenges and opportunities of colonizing other planets and moons.

1. The Challenges of Colonizing Other Planets:

a. Gravity: One of the biggest challenges of colonizing other planets is dealing with differences in gravity. Humans have evolved to live and function under Earth's gravity, and prolonged exposure to other gravity levels can have serious health consequences, including muscle and bone loss, cardiovascular problems, and neurological effects.

b. Atmosphere: Most planets and moons in our solar system have very different atmospheres from Earth, and some have no atmosphere at all. This means that colonists would need to bring their own air supply, and create a habitable atmosphere within their living spaces.

c. Temperature: Temperatures on other planets and moons can vary widely, from scorching hot to freezing cold. Maintaining a stable temperature for human habitation would require advanced heating and cooling systems.

d. Radiation: Space is full of dangerous radiation that can be harmful to humans, and planets without a protective atmosphere or magnetic field are even more vulnerable. Colonists would need to find ways to shield themselves from this radiation.

e. Resources: Establishing colonies on other planets would require a significant amount of resources, including water, food, building materials, and energy sources. These resources would either need to be brought from Earth or obtained from the planet or moon itself, which would require advanced technology and infrastructure.

2. Opportunities of Colonizing Other Planets:

a. Exploration and Discovery: Colonizing other planets would allow us to explore and learn more about the universe and our place within it. We would be able to study the geology, climate, and atmosphere of other planets in much greater detail than we can with robotic missions.

b. Resource Utilization: Many planets and moons contain valuable resources that could be utilized to benefit humanity. For example, Mars has abundant supplies of water, which could be used to create fuel and sustain human life.

c. Scientific Research: Colonizing other planets would provide opportunities for scientific research in a wide range of fields, including biology, geology, physics, and astronomy.

d. International Cooperation: The challenges of colonizing other planets would require cooperation and collaboration among nations and organizations around the world. This could lead to greater international cooperation and understanding.

3. Current and Future Efforts to Colonize Other Planets:

a. Mars: Mars is currently the most viable candidate for human colonization, due to its relative proximity and similarities to Earth. Multiple organizations and space agencies are currently working on plans to send humans to Mars and establish a permanent colony.

b. Moon: The Moon has also been proposed as a potential site for human colonization, due to its proximity and potential resources. NASA has plans to establish a lunar base by the 2030s.

c. Other Planets and Moons: Other planets and moons, such as Venus, Europa, and Titan, have also been proposed as potential sites for human colonization, but face significant challenges due to their extreme environments.

In conclusion, colonizing other planets is a complex and challenging task that requires significant technological advancements, resources, and cooperation between nations. However, the potential benefits, such as expanding the human presence in space, finding new resources, and mitigating the risks of natural disasters on Earth, are undeniable.

As we continue to explore and learn more about our solar system and beyond, it is important to consider the ethical implications of colonizing other worlds and to prioritize the preservation of the natural environments we encounter. We must also ensure that any efforts to colonize other planets are carried out responsibly, sustainably, and with a long-term perspective in mind.

As we move forward, the field of astrobiology will undoubtedly play a crucial role in informing our understanding of the potential habitability of other worlds and the implications of human intervention. While we have much to learn, the pursuit of knowledge and the exploration of new frontiers are fundamental to human curiosity and progress.

Chapter 42: Space Exploration - The history of human spaceflight and the future of space exploration.

Since the first successful launch of the Soviet satellite Sputnik 1 in 1957, human space exploration has captured the imagination of people around the world. In the decades since, humans have ventured further and further into space, with many historic achievements and setbacks along the way. In this chapter, we will take a closer look at the history of human spaceflight, the current state of space exploration, and what the future may hold.

The Early Years

The first human to venture into space was Soviet cosmonaut Yuri Gagarin, who completed a single orbit of the Earth aboard the Vostok 1 spacecraft on April 12, 1961. This achievement was followed by a series of increasingly ambitious missions by both the United States and the Soviet Union, including the famous Apollo program, which put humans on the Moon for the first time in 1969.

During this early period of space exploration, there were many technological challenges to overcome, including developing reliable spacecraft, building suitable launch vehicles, and figuring out how to keep humans alive in the harsh environment of space. Many of these challenges were met with great success, but there were also many setbacks, including the tragic loss of several astronauts in accidents such as the 1986 Challenger disaster.

The International Space Station

In 1998, the International Space Station (ISS) was launched, marking a new era in human space exploration. The ISS is a joint project between the United States, Russia, Europe, Japan, and Canada, and serves as a research platform for scientists from around the world. It has been continuously inhabited since November 2000, and is the largest human-made object in space, with a mass of over 400,000 kg.

The ISS has been the site of many groundbreaking experiments and discoveries, including studies of the effects of long-term spaceflight on the human body, investigations of the behavior of materials in microgravity, and research into the origins of the universe. It has also served as a test bed for new technologies that will be needed for future missions to explore the solar system and beyond.

Exploring the Solar System

While the ISS is focused on research in low Earth orbit, there have also been many missions to explore the solar system and beyond. Some of the most notable of these missions include:

- **The Voyager missions:** Launched in 1977, the Voyager 1 and Voyager 2 spacecraft were designed to study the outer planets of the solar system. They made many groundbreaking discoveries, including the first close-up images of Jupiter, Saturn, Uranus, and Neptune, and are currently the most distant human-made objects from Earth.

- **The Mars rovers:** Since the landing of the Sojourner rover in 1997, there have been several successful missions to explore the surface of Mars. The most recent of these is the Perseverance rover, which landed in February 2021 and is equipped with advanced scientific instruments to search for signs of past or present life on the Red Planet.
- **The New Horizons mission:** Launched in 2006, the New Horizons spacecraft made a historic flyby of the dwarf planet Pluto in 2015, providing the first close-up images of this distant world. It is currently on its way to explore other objects in the Kuiper Belt, the region of the solar system beyond Neptune.
- **The Cassini-Huygens mission:** Launched in 1997, the Cassini spacecraft spent 13 years orbiting Saturn and studying its many moons. It also carried the Huygens probe, which successfully landed on the surface of the moon Titan in 2005, providing the first close-up images of this mysterious world.

The Future of Space Exploration

The future of space exploration is exciting and filled with possibilities. In the coming years and decades, humans will continue to explore our solar system and beyond. There are many missions and projects already in the planning stages, and others that are still just ideas.

One of the most exciting missions in the works is NASA's Artemis program, which aims to land humans on the Moon again by 2024. This time, the plan is to establish a sustainable presence on the Moon, with the eventual goal of using it as a stepping stone to Mars. The Artemis program will also focus on studying the lunar surface and developing technologies to support human exploration.

In addition to the Moon, Mars remains a key target for exploration. Several missions are already in progress, including NASA's Perseverance rover, which landed on Mars in February 2021, and the European Space Agency's ExoMars mission, which is set to launch in 2022. Both missions aim to search for signs of past or present life on the red planet.

Beyond Mars, there are many other destinations that could be explored in the coming years. For example, NASA is planning a mission to visit Europa, one of Jupiter's moons that is believed to have a subsurface ocean. This mission could search for signs of life in the ocean or explore the potential for humans to eventually colonize the moon.

Private companies are also getting involved in space exploration, with SpaceX and Blue Origin both working on plans for commercial spaceflight and colonization. SpaceX has already sent astronauts to the International Space Station and is developing its Starship spacecraft for missions to the Moon, Mars, and beyond.

The future of space exploration will also depend on advances in technology, such as improved propulsion systems, better life support systems, and more advanced robotics. As technology continues to improve, it will open up new possibilities for exploration and colonization.

Overall, the future of space exploration is bright and full of possibilities. With continued investment and advances in technology, humans will continue to push the boundaries of what is possible and expand our understanding of the universe.

Chapter 43: Spacecraft - The different types of spacecraft used for exploring our solar system and beyond

Spacecraft are vehicles designed to operate beyond the Earth's atmosphere and explore space. Over the years, humans have developed various types of spacecraft, each with a unique set of capabilities and functions. Spacecraft have played a crucial role in exploring our solar system, studying the universe, and advancing our understanding of space.

1. Unmanned Spacecraft

Unmanned spacecraft, as the name suggests, do not have any crew on board. They are primarily used for scientific exploration and data collection. The unmanned spacecraft can be divided into two categories:

- **Flyby spacecraft:** These spacecraft fly by planets, moons, asteroids, or comets to study them from a distance. Examples of flyby spacecraft include Pioneer, Voyager, and New Horizons.
- **Orbital spacecraft:** These spacecraft orbit planets, moons, asteroids, or comets to study them in detail. Examples of orbital spacecraft include the Mars Reconnaissance Orbiter and the Lunar Reconnaissance Orbiter.

2. Manned Spacecraft

Manned spacecraft, also known as crewed spacecraft, carry human beings into space. These spacecraft are designed to provide a safe and comfortable environment for humans to live and work in space. Manned spacecraft can be divided into two categories:

- **Suborbital spacecraft:** These spacecraft travel to space but do not enter orbit around the Earth. Examples of suborbital spacecraft include the X-15, SpaceShipOne, and Virgin Galactic's SpaceShipTwo.
- **Orbital spacecraft:** These spacecraft enter into orbit around the Earth and can remain in space for extended periods. Examples of orbital spacecraft include the International Space Station, the Space Shuttle, and the Soyuz spacecraft.

3. Probes

Probes are unmanned spacecraft designed to explore space beyond our solar system. These spacecraft are equipped with scientific instruments and sensors that collect data and send it back to Earth. Probes can be divided into two categories:

- **Flyby probes:** These probes fly by planets, moons, asteroids, or comets to study them from a distance. Examples of flyby probes include Voyager 1 and 2, and New Horizons.

- **Orbital probes:** These probes orbit planets, moons, asteroids, or comets to study them in detail. Examples of orbital probes include the Mars Reconnaissance Orbiter and the Lunar Reconnaissance Orbiter.

4. Landers and Rovers

Lander and rovers are unmanned spacecraft designed to land on the surface of planets, moons, asteroids, or comets to study them in detail. Landers are stationary spacecraft that remain in one place and perform scientific experiments, while rovers are mobile spacecraft that can move around and explore their surroundings. Examples of landers and rovers include the Viking landers on Mars, the Spirit and Opportunity rovers on Mars, and the Philae lander on the comet 67P/Churyumov–Gerasimenko.

5. Sample Return Missions

Sample return missions are unmanned spacecraft that travel to planets, moons, asteroids, or comets, collect samples, and return them to Earth for analysis. These missions provide scientists with a unique opportunity to study the composition and history of celestial bodies in detail. Examples of sample return missions include the Stardust mission, which collected samples from a comet, and the Hayabusa and Hayabusa2 missions, which collected samples from asteroids.

In conclusion, spacecraft are essential tools for exploring space and advancing our understanding of the universe. With the development of new technologies, we can expect to see more advanced spacecraft in the future, capable of exploring deeper into space and helping us answer some of the most fundamental questions about the cosmos.

Chapter 44: Space Missions - A tour of some of the most significant space missions in history.

Humans have been fascinated with space since the dawn of time, and we've come a long way since the first primitive attempts to explore the cosmos. Today, we have sent spacecraft to every planet in our solar system and beyond, revealing stunning discoveries and answering some of the most profound questions about the universe. In this chapter, we will take a tour of some of the most significant space missions in history and see what they have taught us about our place in the cosmos.

1. Apollo 11 (1969)

The Apollo 11 mission was the first crewed mission to land on the Moon. Astronauts Neil Armstrong and Edwin "Buzz" Aldrin became the first humans to set foot on another celestial body, while Michael Collins orbited above. The mission was a triumph of human ingenuity and a defining moment in our exploration of space.

2. Voyager 1 and 2 (1977)

The twin Voyager spacecraft were launched in 1977 to study the outer planets of our solar system. Voyager 1 flew past Jupiter and Saturn, while Voyager 2 continued on to Uranus and Neptune. Both spacecraft are still sending data back to Earth as they journey into interstellar space, making them the farthest man made objects from Earth.

3. <u>Hubble Space Telescope (1990)</u>

The Hubble Space Telescope is a large, space-based observatory that has revolutionized our understanding of the universe. It has captured stunning images of galaxies, nebulae, and other cosmic objects, as well as helped us to measure the size and age of the universe.

4. <u>Mars Pathfinder (1996)</u>

The Mars Pathfinder mission was the first successful mission to land on Mars since the Viking landers in the 1970s. The mission was designed to study the geology and atmosphere of Mars and featured the Sojourner rover, the first rover to operate on another planet.

5. <u>Cassini-Huygens (1997)</u>

The Cassini-Huygens mission was a joint project between NASA, the European Space Agency (ESA), and the Italian Space Agency (ASI). The spacecraft was launched in 1997 and arrived at Saturn in 2004. It studied the planet and its many moons, including Titan, the only moon in our solar system with a substantial atmosphere.

6. <u>New Horizons (2006)</u>

The New Horizons spacecraft was launched in 2006 with the goal of studying Pluto and its moons. The spacecraft arrived at Pluto in 2015, providing our first close-up images of the dwarf planet and revealing a complex and fascinating world.

7. <u>Kepler (2009)</u>

The Kepler spacecraft was designed to search for exoplanets, or planets that orbit stars other than our Sun. It monitored the brightness of over 100,000 stars in a single patch of the sky, looking for the telltale dip in brightness that occurs when a planet passes in front of its host star. Kepler discovered thousands of exoplanets, greatly expanding our understanding of the diversity of planets in our galaxy.

8. <u>Curiosity (2011)</u>

The Curiosity rover was the largest and most advanced rover ever sent to Mars. It landed on the planet in 2012 and has been exploring the Gale Crater ever since, studying the geology and searching for signs of past or present microbial life.

9. <u>James Webb Space Telescope (2021)</u>

The James Webb Space Telescope is a successor to the Hubble Space Telescope and will be the largest and most powerful space telescope ever launched. It will study the universe in the infrared portion of the spectrum, allowing it to see through dust clouds and observe some of the earliest galaxies in the universe.

In conclusion, space missions have played a crucial role in advancing our understanding of the universe and our place in it. From the first human spaceflight to the exploration of other planets and moons in our solar system, these missions have expanded our knowledge and opened up new avenues of research. They have also provided inspiration for future generations of scientists and explorers. As we continue to push the boundaries of space exploration, it is important to remember the accomplishments and sacrifices of those who came before us and to continue to work towards new discoveries and innovations.

Chapter 45: Space Agencies - An overview of the major space agencies around the world.

Space exploration is a collaborative effort that involves the contributions of many nations and organizations around the world. In this chapter, we will take a look at some of the major space agencies and their roles in advancing our understanding of the cosmos.

1. National Aeronautics and Space Administration (NASA):

The National Aeronautics and Space Administration (NASA) is the space agency of the United States. Founded in 1958, NASA has been responsible for many groundbreaking space missions, including the Apollo moon landings, the Space Shuttle program, and the Hubble Space Telescope. NASA is also actively involved in current and future missions to explore our solar system and beyond, such as the Mars rovers, the James Webb Space Telescope, and the upcoming Artemis program, which aims to land the first woman and the next man on the Moon by 2024.

2. European Space Agency (ESA):

The European Space Agency (ESA) is a collaborative effort between 22 European countries. Established in 1975, the agency's main objective is to carry out research and development in the field of space science and technology. ESA is responsible for a number of important space missions, including the Rosetta spacecraft, which made history by landing on a comet in 2014, and the Gaia mission, which is mapping the Milky Way galaxy in unprecedented detail.

3. Russian Space Agency (Roscosmos):

The Russian Space Agency, also known as Roscosmos, is the space agency of Russia. Established in 1992 after the dissolution of the Soviet Union, Roscosmos has been responsible for a number of significant space missions, including the launch of the first artificial satellite, Sputnik 1, in 1957. More recently, Roscosmos has been involved in missions to the International Space Station and the exploration of Mars.

4. China National Space Administration (CNSA):

The China National Space Administration (CNSA) is the space agency of China. Established in 1993, CNSA has made significant progress in the field of space exploration in recent years, including the successful landing of the Chang'e 4 spacecraft on the far side of the Moon in 2019. CNSA has also expressed its intention to establish a permanent human presence on the Moon and eventually send astronauts to Mars.

5. Japan Aerospace Exploration Agency (JAXA):

The Japan Aerospace Exploration Agency (JAXA) is the space agency of Japan. Established in 2003, JAXA has been involved in a number of important space missions, including the

Hayabusa spacecraft, which successfully returned samples from an asteroid in 2010. JAXA is also involved in missions to study Earth's environment, such as the Global Precipitation Measurement mission and the Greenhouse Gases Observing Satellite.

6. <u>Indian Space Research Organisation (ISRO):</u>

The Indian Space Research Organisation (ISRO) is the space agency of India. Established in 1969, ISRO has been responsible for a number of significant space missions, including the Mars Orbiter Mission in 2014, which made India the first country to successfully reach Mars on its first attempt. ISRO is also involved in missions to study Earth's environment, such as the Oceansat-3 satellite and the Chandrayaan-2 mission to the Moon.

7. <u>Canadian Space Agency (CSA):</u>

The Canadian Space Agency (CSA) is the national space agency of Canada. It was established in 1989, and since then it has played a vital role in the development of space technology and exploration. The CSA is responsible for managing all of Canada's space-related activities, including the development of spacecraft and satellites, the coordination of space missions, and the training of astronauts.

One of the most significant contributions of the CSA has been the development of the Canadarm, a robotic arm that has been used extensively on the Space Shuttle and the International Space Station. The Canadarm has been instrumental in a wide range of activities, from the deployment and repair of satellites to the construction of the International Space Station.

The CSA has also been involved in a number of space missions, both in collaboration with other space agencies and on its own. In 2013, the CSA launched the Cassiope satellite, which is used for scientific research and for providing communication services to remote areas of Canada. The agency has also been involved in several Mars missions, including the Phoenix lander in 2007 and the Mars Science Laboratory (Curiosity) rover in 2011.

In addition to its involvement in space missions and technology development, the CSA is also focused on promoting education and public awareness of space-related topics. The agency offers a range of educational programs and resources for students, as well as public outreach events and activities.

Overall, the Canadian Space Agency has made significant contributions to space exploration and technology development, and it continues to play an important role in the global space community.

In conclusion, space exploration is a fascinating and complex field that has captured the imagination of people all around the world. From the first steps on the Moon to the discovery of exoplanets, humans have made incredible strides in our understanding of the universe and our place in it. The work of space agencies and their missions have brought about numerous technological advancements that have also had an impact on daily life.

With the increasing interest and investment in space exploration, we can expect to see even more exciting discoveries and achievements in the future. International cooperation and collaboration will play a critical role in the continued exploration of space and the development of new technologies. As we look to the future, it is clear that space exploration will remain a crucial field of study that will continue to inspire us and expand our understanding of the universe.

Chapter 46: Space Technology

Space exploration has always been a catalyst for the development of new and innovative technologies. From the early days of rocketry and satellite communications to the advanced robotic probes and human spaceflight programs of today, space technology has pushed the boundaries of what is possible and opened up new frontiers of scientific discovery.

In this chapter, we will take a look at some of the key technologies that have made space exploration possible and their impact on our daily lives.

Rocket Technology:

One of the most critical technologies for space exploration is rocket technology. Rockets provide the thrust necessary to launch spacecraft and send them on their way to explore the solar system and beyond. The development of powerful and reliable rockets was essential for the success of the Apollo missions, which put humans on the moon for the first time.

Today, rockets continue to play a crucial role in space exploration, with companies like SpaceX and Blue Origin leading the way in developing new and innovative rocket designs that are more efficient, reusable, and cost-effective than ever before.

Satellite Technology:

Satellites are another essential technology for space exploration. They allow us to communicate across vast distances, observe the Earth from space, and gather data about the universe beyond. The first artificial satellite, Sputnik 1, was launched by the Soviet Union in 1957, and since then, thousands of satellites have been sent into orbit around the Earth.

Today, satellites are used for a wide range of applications, from GPS navigation and weather forecasting to military reconnaissance and scientific research. They have revolutionized the way we communicate, gather information, and understand our planet and the universe.

Robotics Technology:

Robotic probes have enabled us to explore the most distant reaches of our solar system and beyond. From the first robotic probes to the moon to the recent missions to Mars and beyond, robotic technology has allowed us to gather data and images from places that would otherwise be impossible to reach.

Robotic technology has also found many practical applications on Earth, from medical robots that perform complex surgeries to industrial robots that automate manufacturing processes. The development of new and more advanced robotics technology will continue to play a significant role in both space exploration and everyday life.

Materials Science:

Space exploration has also driven advances in materials science, from lightweight and durable materials for spacecraft to advanced composites for space suits and other equipment. These materials have applications beyond space exploration, from aerospace to automotive and construction industries.

Spinoff Technologies:

One of the most significant benefits of space exploration has been the spinoff technologies that have emerged from space technology research. Technologies developed for space exploration have led to breakthroughs in a variety of fields, including medicine, transportation, and energy production.

For example, research on water recycling systems for space missions has led to the development of efficient water treatment systems for use in homes and businesses. Other spinoff technologies include energy-efficient insulation, solar panels, and lightweight materials.

Conclusion:

Space technology has made it possible for us to explore the universe and understand our place in it. The development of new and innovative technologies has enabled us to achieve unprecedented levels of success in space exploration, from landing humans on the moon to exploring the far reaches of our solar system.

At the same time, space technology has had a profound impact on our daily lives, from GPS navigation and weather forecasting to medical robotics and energy-efficient materials. As we continue to push the boundaries of what is possible in space exploration, we can look forward to even more breakthroughs and spinoff technologies that will benefit humanity for years to come.

Chapter 47: Space Debris - The Growing Problem of Debris in Earth's Orbit and Its Impact on Space Exploration

As humans continue to explore space, we are leaving behind a trail of debris that is posing a growing threat to our future missions. This debris, also known as space junk, includes everything from old satellites and rocket stages to broken pieces of hardware and even tiny fragments of paint. With thousands of pieces of debris currently orbiting Earth at speeds of up to 17,500 miles per hour, collisions with functioning spacecraft and satellites are becoming more common, and the risk of catastrophic damage is increasing.

In this chapter, we will explore the problem of space debris, its causes, and the potential solutions for mitigating its impact on space exploration.

Causes of Space Debris

The primary cause of space debris is human activity in space. Since the launch of the first artificial satellite, Sputnik 1, in 1957, humans have sent thousands of spacecraft and satellites into Earth's orbit. While many of these objects have been intentionally deorbited or placed in graveyard orbits, many others remain in orbit as space debris.

The other major cause of space debris is the natural process of collisions between objects in space. Even small pieces of debris can cause significant damage to spacecraft and satellites due to the high speeds at which they travel.

Impact of Space Debris on Space Exploration

The presence of space debris poses a significant risk to space exploration. When a piece of debris collides with a functioning spacecraft or satellite, it can cause severe damage or even total destruction. This can result in mission failure, loss of communication, and even endanger human life.

Space debris also limits our ability to launch new missions into space. As the amount of debris in Earth's orbit continues to grow, the risk of collisions increases, making it more challenging to find safe launch windows and orbital paths. This, in turn, limits our ability to explore space and gather valuable scientific data.

Potential Solutions for Space Debris Mitigation

Several potential solutions have been proposed to mitigate the impact of space debris on space exploration. These include:

1. **Active Debris Removal:** This involves sending spacecraft into orbit to collect and remove debris. This method has the potential to reduce the amount of debris in orbit significantly. However, it is expensive and technically challenging.
2. **Passive Debris Removal:** This involves designing spacecraft and satellites with materials that will burn up

upon re-entry into the Earth's atmosphere at the end of their operational life. This method has the advantage of being more cost-effective and easier to implement.

3. **Collision Avoidance:** This involves tracking and monitoring the location of space debris and adjusting the trajectory of functioning spacecraft and satellites to avoid collisions. This method is already in use but requires constant monitoring and adjustment.

4. **International Cooperation:** The problem of space debris is a global issue that requires international cooperation. Governments and space agencies around the world must work together to develop and implement effective solutions to reduce the amount of debris in orbit and prevent future debris.

Conclusion

Space debris is a growing problem that poses a significant risk to space exploration. As we continue to send more spacecraft and satellites into orbit, the amount of debris will only continue to increase. Mitigating the impact of space debris requires a concerted effort from governments, space agencies, and private companies around the world. By developing and implementing effective solutions, we can reduce the risk of collisions and ensure a safer and more sustainable future for space exploration.

Chapter 48: Astronomy Outreach

Astronomy is a fascinating and inspiring field of study, but for many people, it can also be intimidating and difficult to understand. Astronomy outreach seeks to bridge this gap by bringing astronomy to the public in accessible and engaging ways. From amateur astronomy clubs to planetarium shows and public lectures, astronomy outreach has become an essential part of the scientific community's effort to share the wonders of the universe with the public.

Importance of Astronomy Outreach

Astronomy outreach serves several important purposes. First, it helps to inspire the next generation of scientists and astronomers. By exposing young people to the wonders of the universe, astronomy outreach programs can encourage them to pursue careers in science, technology, engineering, and mathematics (STEM) fields.

Second, astronomy outreach helps to promote scientific literacy and critical thinking skills. By explaining complex scientific concepts in easy-to-understand terms, astronomy outreach programs can help people better understand and appreciate the scientific method, and can encourage them to think critically about scientific claims and evidence.

Finally, astronomy outreach can help to foster a sense of community and shared interest among people from diverse backgrounds. Astronomy is a subject that can appeal to people of all ages, genders, races, and nationalities, and astronomy outreach programs can provide a space for people to come together and share their passion for the universe.

Types of Astronomy Outreach

There are many different types of astronomy outreach programs, ranging from informal activities like stargazing parties to formal educational programs offered by schools and museums. Here are some of the most common types of astronomy outreach programs:

1. **Amateur Astronomy Clubs:** Amateur astronomy clubs are groups of amateur astronomers who come together to share their knowledge and love of astronomy. These clubs often organize stargazing events and other astronomy-related activities, and can be a great way for people to learn more about astronomy in a supportive and welcoming environment.
2. **Planetarium Shows:** Planetariums are special theaters that simulate the night sky on a dome-shaped screen. Planetarium shows can be a great way to introduce people to the night sky and to teach them about astronomy in an immersive and engaging way.
3. **Public Lectures:** Many universities, museums, and other organizations offer public lectures on astronomy and related topics. These lectures can be a great way to learn

about the latest discoveries in astronomy and to hear from experts in the field.

4. **Outreach Programs for Schools:** Many schools offer astronomy outreach programs that bring telescopes and other equipment to schools to give students hands-on experience with astronomy. These programs can be a great way to get young people excited about science and to encourage them to pursue STEM careers.

5. **Citizen Science Projects:** Citizen science projects allow members of the public to contribute to scientific research by collecting data or analyzing data from telescopes or other instruments. These projects can be a great way for people to learn about astronomy while also contributing to important scientific research.

Challenges in Astronomy Outreach

Despite the many benefits of astronomy outreach, there are also several challenges that must be addressed. One of the biggest challenges is funding, as many outreach programs rely on grants and donations to operate. Additionally, many outreach programs struggle to reach underrepresented communities, including people of color, low-income families, and people living in rural areas.

Another challenge in astronomy outreach is the need for effective communication. Astronomical concepts can be difficult to understand, and it can be challenging to explain complex ideas in a way that is accessible to a broad audience. Effective communication is essential for successful outreach programs, and many organizations invest in training their staff and volunteers in effective communication strategies.

Finally, astronomy outreach programs must also navigate the challenges posed by the COVID-19 pandemic. Many outreach programs have had to shift to virtual formats, which can make it more difficult to engage with audiences and to provide hands-on experiences with astronomy.

In conclusion, astronomy outreach plays a critical role in educating the public about the wonders of the universe and inspiring future generations to pursue careers in science. The diverse range of outreach programs, from planetarium shows and public lectures to citizen science projects and social media campaigns, provides ample opportunities for people of all ages and backgrounds to engage with astronomy. By sharing our knowledge and enthusiasm for the cosmos, we can foster a greater appreciation for science and promote scientific literacy in society. It is imperative that we continue to support and expand astronomy outreach efforts to ensure that everyone has the opportunity to explore and appreciate the wonders of the universe.

Chapter 49: Amateur Astronomy - The many ways amateurs can contribute to astronomy and space science.

Astronomy has been a popular hobby for centuries, with many amateurs dedicating their time and resources to observing the night sky. With the increasing availability of affordable equipment and technology, amateur astronomy has become more accessible than ever before. In addition to enjoying the wonders of the universe, amateur astronomers can also make significant contributions to science by helping to discover new objects and phenomena, collecting data, and sharing their knowledge with others.

1. History of Amateur Astronomy

Amateur astronomy has a long and rich history dating back to ancient times when early civilizations observed the stars and planets. During the Renaissance, astronomers such as Galileo Galilei and Johannes Kepler used telescopes to make groundbreaking discoveries. In the 19th century, the popularity of astronomy as a hobby grew, with many enthusiasts building their own telescopes and making observations. Today, amateur astronomers continue to make important contributions to the field of astronomy.

2. Types of Amateur Astronomy

3. **There are many different types of amateur astronomy**, including visual observing, astrophotography, and citizen science projects. Visual observing involves using a telescope or binoculars to observe the night sky and record observations of celestial objects. Astrophotography involves using a camera to take images of the night sky. Citizen science projects involve collaborating with professional scientists to collect data and contribute to scientific research.

4. Contributions to Astronomy

Amateur astronomers have made significant contributions to astronomy throughout history. In the early 20th century, amateur astronomers discovered a number of comets and asteroids. More recently, amateur astronomers have helped to discover new exoplanets, supernovae, and other celestial objects. Citizen science projects have also played a significant role in modern astronomy, with volunteers contributing to research in areas such as exoplanet detection and galaxy classification.

5. Equipment and Technology

Modern technology has made amateur astronomy more accessible than ever before. Telescopes, cameras, and other equipment can be purchased at affordable prices, and software programs can be used to aid in observing and data analysis. Amateur astronomers also have access to online resources and communities, where they can share their knowledge, ask questions, and collaborate with other enthusiasts.

6. Challenges and Considerations

While amateur astronomy can be a rewarding and exciting hobby, there are also some challenges and considerations to keep in mind. Light pollution in urban areas can make observing difficult, and weather conditions can also impact observations. It is important for amateur astronomers to prioritize safety and follow proper procedures when using equipment. Additionally, it is important to respect intellectual property rights and obtain proper permissions when sharing images or data.

7. Conclusion

Amateur astronomy is a vibrant and important part of the astronomy community. Enthusiasts of all ages and experience levels can enjoy the wonders of the universe and make important contributions to science through observing, collecting data, and participating in citizen science projects. With the increasing availability of affordable equipment and technology, amateur astronomy is more accessible than ever before, and is sure to continue to inspire and engage future generations.

Chapter 50: The Future of Astronomy

Astronomy has come a long way since the earliest observations of the stars and planets. From the invention of the telescope to the discovery of black holes and gravitational waves, we have made incredible progress in our understanding of the universe. But there is still so much to discover and explore, and the future of astronomy is full of exciting possibilities.

In this chapter, we will explore some of the major areas of research that are shaping the future of astronomy, including new telescopes and observatories, space missions, and emerging technologies.

New Telescopes and Observatories

One of the most exciting areas of astronomy is the development of new telescopes and observatories that will allow us to see deeper into the universe than ever before. Here are a few examples:

1. **The James Webb Space Telescope (JWST):** This next-generation space telescope, set to launch in 2021, will be the largest and most powerful ever built. It will observe in the infrared spectrum and will be capable of seeing through dust clouds and detecting the faintest objects in the universe. The JWST will revolutionize our understanding of the early universe, exoplanets, and the formation of galaxies.

2. **The Square Kilometer Array (SKA):** The SKA is a radio telescope currently being built in Australia and South Africa. When completed, it will be the largest and most sensitive radio telescope in the world, capable of detecting faint signals from the earliest days of the universe. The SKA will help us better understand the formation and evolution of galaxies, the nature of dark matter and dark energy, and the origins of life in the universe.

3. **The Large Synoptic Survey Telescope (LSST):** This ground-based telescope, set to begin operations in 2023, will survey the entire southern sky every few nights, creating the most comprehensive map of the universe ever made. The LSST will detect billions of objects, including new planets, supernovae, and asteroids, and will help us better understand the nature of dark matter and dark energy.

Space Missions

Space missions are another important area of astronomy research. Here are a few upcoming missions that are set to make major contributions to our understanding of the universe:

1. **Europa Clipper:** This NASA mission, set to launch in the mid-2020s, will study Jupiter's moon Europa, which is believed to have a subsurface ocean of liquid water. The mission will search for signs of life and study the moon's geology and potential for future exploration.

2. **Mars Sample Return:** This joint NASA-ESA mission, set to launch in the late 2020s, will collect samples of Martian soil and rocks and return them to Earth for analysis. The samples will provide a wealth of information about the geology and potential habitability of Mars.
3. **WFIRST:** The Wide Field Infrared Survey Telescope (WFIRST) is a NASA mission set to launch in the mid-2020s. It will conduct a survey of the sky in the infrared spectrum, studying dark energy and dark matter, exoplanets, and the early universe.

Emerging Technologies

Finally, emerging technologies are also shaping the future of astronomy. Here are a few examples:

1. **Artificial Intelligence (AI):** AI is becoming increasingly important in astronomy research, as it can analyze vast amounts of data quickly and efficiently. AI is being used to study everything from gravitational waves to exoplanets to the cosmic microwave background radiation.
2. **Quantum Computing:** Quantum computers may be able to solve problems that are currently beyond the capabilities of classical computers. In astronomy, quantum computing could be used to simulate the behavior of large-scale structures in the universe, such as galaxies and clusters of galaxies.

3. **Laser Interferometry:** Laser Interferometry is a technology used to detect gravitational waves. Laser Interferometers use two or more beams of light that are combined to create interference patterns. When a gravitational wave passes through the interferometer, it causes a distortion in space-time, which changes the interference pattern. By measuring these changes, scientists can detect and study gravitational waves.
4. The Laser Interferometer Gravitational-Wave Observatory (LIGO) made the first direct detection of gravitational waves in 2015, which was a major breakthrough in the field of astronomy. Since then, several other detections have been made, including signals from the collision of two neutron stars.
5. Future advancements in laser interferometry technology will allow for even more precise measurements of gravitational waves, which will enable scientists to study the properties of black holes and other extreme cosmic phenomena.
6. Another exciting development in the future of astronomy is the launch of the James Webb Space Telescope (JWST) in 2021. The JWST will be the largest and most powerful space telescope ever built, and will be capable of studying the universe in unprecedented detail. With its infrared capabilities, the JWST will be able to see through dust clouds and observe the earliest galaxies in the universe.

7. In addition, new ground-based telescopes, such as the Extremely Large Telescope (ELT) and the Thirty Meter Telescope (TMT), will have mirrors up to 30 meters in diameter, providing even greater resolution and sensitivity than current telescopes.

8. The search for life beyond Earth will also continue to be a major focus in the future of astronomy. NASA's upcoming missions to Europa and Enceladus, two icy moons in our solar system that may have subsurface oceans, will provide valuable information about the potential for life in our own solar system.

9. Finally, the field of astronomy will continue to benefit from advances in artificial intelligence and machine learning, which will allow for faster and more accurate analysis of large amounts of data. With these tools, scientists will be able to make new discoveries and gain a deeper understanding of the universe.

Recommendations

Thank you for reading this book on astronomy and space exploration! We hope that you have enjoyed learning about the wonders of the universe and the exciting field of space science.

If you are interested in reading more about astronomy and space exploration, here are five book recommendations:

1. **"The Universe in a Nutshell"** by Stephen Hawking - This book provides a fascinating overview of the latest developments in cosmology, from the Big Bang to black holes.
2. **"Cosmos"** by Carl Sagan - This classic book explores the history of the universe and the search for extraterrestrial life.
3. **"The Elegant Universe"** by Brian Greene - This book provides a comprehensive introduction to string theory and the unification of the laws of physics.
4. **"The Martian"** by Andy Weir - This novel tells the story of an astronaut stranded on Mars and his struggle to survive and make it back to Earth.
5. **"Packing for Mars"** by Mary Roach - This book provides a humorous and informative look at the challenges of space exploration, from the physical and psychological effects of long-duration spaceflight to the practicalities of eating, sleeping, and going to the bathroom in space.

We hope that these recommendations inspire you to continue exploring the wonders of the universe and the fascinating field of space science.

www.ingramcontent.com/pod-product-compliance
Lightning Source LLC
Chambersburg PA
CBHW070800220526

45467CB00017B/186